MATLAB® For Engineers Explained

Springer

London
Berlin
Heidelberg
New York
Hong Kong
Milan
Paris
Tokyo

Fredrik Gustafsson and Niclas Bergman

MATLAB® for Engineers Explained

With 123 Figures

 Springer

Fredrik Gustafsson
Department of Electrical Engineering, Linköping University, SE-58183, Linköping, Sweden

Niclas Bergman
SaabTech Systems AB, Data Fusion Group, 175 88 Järfälla, Sweden

British Library Cataloguing in Publication Data
Gustafsson, Fredrik
 MATLAB for Engineers Explained
 1.MATLAB (Computer file) 2.Numerical analysis – Data
 processing 3.Engineering mathematics – Data processing
 I.Title II.Bergman, Niclas
 620'.002855369
 ISBN 1852336978

ISBN 1-85233-697-8 Springer-Verlag London Berlin Heidelberg
a member of BertelsmannSpringer Science+Business Media GmbH
http://www.springer.co.uk

Typesetting: Electronic text files prepared by authors
Printed and bound at The Cromwell Press, Trowbridge, Wiltshire
69/3830-54321 Printed on acid-free paper SPIN 10947050

CONTENTS

GUIDED TOURS

PREFACE

This book is written for students at bachelor and master programs and has four different purposes, which split the book into four parts:

1. To teach first or early year undergraduate engineering students basic knowledge in technical computations and programming using MATLAB. The first part starts from first principles and is therefore well suited both for readers with prior exposure to MATLAB but lacking a solid foundational knowledge of the capabilities of the system and readers not having any previous experience with MATLAB. The foundational knowledge gained from these interactive guided tours of the system will hopefully be sufficient for an effective utilization of MATLAB in the engineering profession, in education and in research.

2. To explain the foundations of more advanced use of MATLAB using the facilities added the last couple of years, such as extended data structures, object orientation and advanced graphics.

3. To give an introduction to the use of MATLAB in typical undergraduate courses in electrical engineering and mathematics, such as calculus, algebra, numerical analysis and statistics. This part also contains introductions and mini-manuals to the most used MATLAB toolboxes. Thus, some chapters require additional MATLAB toolboxes. The idea is to give a brief tutorial on each subject and show the possibilities for applying MATLAB to each application area. We have focused on basic concepts in the applications, without trying to explain all theory behind the examples.

4. The appendix is an extensive reference part to a selection of the most useful matlab functions. The tables summarize complete syntax organized according to theoretical relations, rather than the organization in the MATLAB system.

The book is a beginner's introduction to MATLAB rather than a complete reference to all the thousands of functions available in MATLAB. The goal is to teach a sufficient subset of the functionality and give the reader practical experience on how to find more information.

The second part of the book contains advanced concepts of MATLAB, normally not required in any undergraduate course, but still important for larger projects and thesis work. Among the topics are how to optimize speed of computation, how to construct graphical user interfaces, general data structures and object orientation amongst other things.

The objective of this manuscript is to gradually teach you to use MATLAB. Each chapter starts with a brief description of the content and is followed by a list of MATLAB relevant functions, and some general aspects of the chapter.

A guided tour 1 (Preface)

The core of each chapter consists of one or several guided tours. The idea is that the right column shows what you type in MATLAB and the left column what the purpose is. Consequently, these words explain what you see to the right. There are three different things that may appear to the right:

- files (contained in shadowed boxes),

- MATLAB window prompter and commands typed here (here the function `logo`),

- and all graphics that MATLAB produces from these, just as each plot appears when you copy the text into MATLAB.

File name: `preface.m`

```
% This is the content of
% the file preface.m
```

```
>> logo % This is typed in matlab
```

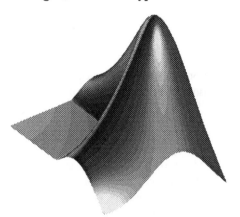

All chapters in the first part of this book are appended by a number of exercises, with solutions in Appendix A. Appendix B contains a reference listing of the presented MATLAB commands. Tables summarizing mathematical areas are found in Appendix C, while Appendix D provides mini-manuals to some common toolboxes. This bibliography contains a subset of English books with a similar scope as this one: an introduction for engineering students.

The most efficient way to learn to use the MATLAB system is to gradually work through each guided tour by copying the code and play around with the results. Curiosity is the mother of understanding, and you should therefore improvise and alter the input somewhat and try your own ideas of how to use the MATLAB system and its functions. Much complementary material and examples in the book are available to down-load from the book home page `http://www.control.isy.liu.se/books/matlab`, where also a self examination test is available, where you can compare your performance to other students!

Finally, we gratefully acknowledge all kind of valuable feedback, positive as negative, we have got from the 2326 students at Linköping University who have passed the course so far! We also thank all teachers involved in teaching the course, colleagues helping us with Latex, HTML and Perl programming, students and teachers from Lund Institute of Technology, Chalmers and the Royal Institute of Technology.

Part 1

LEARNING MATLAB

1 Introduction

The name MATLAB is an acronym that stands for MATrix LABoratory. MATLAB is an interactive environment for performing technical computations. The software package MATLAB has been commercially available since 1984 and is now considered a standard tool at most universities and many industries worldwide. The core of MATLAB consists of compiled C-coded routines for matrix computation, numerical analysis and graphics. However, the major functionality in MATLAB is coded as so called *m-files*, which are plain text files having file extension .m. These files consist of source code written in the MATLAB programming language. All m-files can be viewed and edited in any standard text editor. A user may therefore copy and edit any of the m-files distributed with the MATLAB software package, or the user may write new m-files from scratch, adding extra functionality to the interactive computation environment of MATLAB.

The compiled functions built into the MATLAB kernel together with the standard m-files provide the user of MATLAB with access to many powerful routines for numerical matrix algebra. Specific application oriented m-files are collected in packages referred to as a *toolbox*, often having an application specific notation familiar to researchers and engineers active in the field. There are both commercial and free toolboxes that add powerful functionality to MATLAB from a specific application field. There are toolboxes for signal processing, simulation, symbolic computation, control theory, optimization, system identification and several other fields of applied science and engineering.

Platforms and software versions

MATLAB is available for several different *platforms*. These include PC with Windows or Linux, Macintosh and Sun workstations. Once inside MATLAB, virtually all syntax is platform independent and m-files written for MATLAB on one platform is generally executable under all other platforms. The only exception to this rule is m-files that perform hardware interface operations or specific operation system calls, and m-files that have been compiled, also called *mex-files*. This book has been developed on different Windows and Sun Solaris platforms, using MATLAB versions ranging from 5.2 to 6.5.

The manufacturer of MATLAB, the Mathworks Inc., uses a dual numbering of MATLAB with versions and releases. A release is basically a running number of the distribution CD's containing MATLAB and all of its toolboxes with different version numbers. MATLAB version 5.2 is part of Release 10, MATLAB version 5.3 is part of Release 11 and MATLAB version 6.0 is part of Release 12, respectively. At the time this book goes into print the current version of MATLAB is version 6.5 which is part of Release 13. The release versions will in the sequel be abbreviated R13 and so on. Note, however, that most of the material is timeless, and only major differences will be pointed out.

Preliminaries

Content: Starting MATLAB and searching for on-line help.
Functions:
`help, heldesk, demo, type, tour, ver`

On PC with Windows and on Mac systems, MATLAB is started from the graphical user interface of the operating system and it should be obvious where to find MATLAB. For example, using Windows, a MATLAB folder should be available under the start menu after successful installation of the system. On most Unix systems, MATLAB is launched from a terminal window by executing the command `matlab`.

A guided tour 2 (Starting MATLAB *)*

When launching MATLAB an initial introduction message, like the one shown to the right, provides the user with some suggestions on where to find help and information. Try typing the commands `demo` and `tour` to get a brief introduction to the MATLAB system.

```
To get started, type one of these:
helpwin, helpdesk, or demo.
For product information, type
tour or visit www.mathworks.com.
```

The version of MATLAB, SIMULINK and all installed toolboxes is available from the output of the command `ver`.

```
>> ver matlab
MATLAB Version 6.1.0 on PCWIN
MATLAB License Number: DEMO
MATLAB Toolbox Version 6.1
```

Terminate MATLAB with the command `quit` or `exit`.

```
>> quit
```

There is an advanced system of *documentation* and *help information* available in MATLAB. The command `helpdesk` loads the main MATLAB Help Desk page into a browser. This hyper linked help system contains comprehensive manuals and tutorials together with detailed searchable information about all commands in the installed MATLAB system. You are encouraged to get acquainted with the MATLAB Help Desk since it will prove very useful when troubleshooting in MATLAB. The Help Desk is convenient since you can search for the name of functions with some desirable functionality. The command `help` *function* prints the help text for a specific function directly in the command window. This is often a good way to

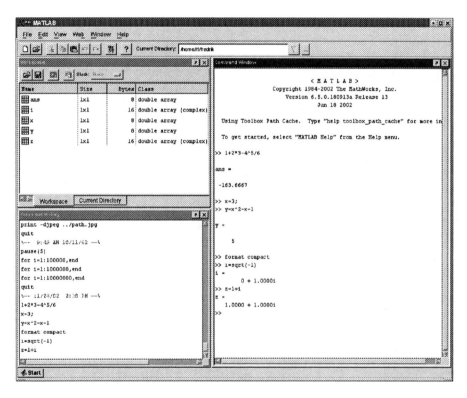

Figure 1.1 The graphical interface to the MATLAB workspace is provided in Release 10 and higher. It provides a number of easy to use tools for interactively changing the settings of your MATLAB system, editing the workspace variables, moving around in the directory structure and accessing previously typed MATLAB syntax.

get quick information about the syntax and functionality of a MATLAB function that you know by name.

An important feature with MATLAB is that the source code for most of the functions is available to browse, copy and alter. Some of the functions in MATLAB are compiled from source C-code and built into the MATLAB kernel, but most of the functions are m-files that have public source code and can be listed using the command **type** *function*.

A guided tour 3 (Preliminaries)

The function **sum** is part of the MATLAB kernel, while the matrix trace function **trace** is an m-file, which is open source but in turn uses the built-in function **sum**.

Note that the m-file contains three parts. The first row is the function header defining the syntax, the first paragraph of comments starting with % provides a brief help text, and the last row is the actual command sequence that executes the m-file. The first paragraph of help information is displayed when typing **help** followed by the function name. Details regarding the matlab *search path* system and issues of writing your own m-files are discussed in great detail in Section 8.

You can open any m-file in the built-in editor by typing **edit**, or in a standard text editor. The search path to the m-file is printed using the command **which**. Note that we will use three dots whenever a line break is needed to fit the page format. This convention is also used in MATLAB, see page 23.

```
>> type sum
sum is a built-in function.
>> type trace

function t = trace(a)
%TRACE  Sum of diagonal elements.
%    TRACE(A) is the sum of the
%    diagonal elements of A,
%    which is also the sum of the
%    eigenvalues of A.

%    The MathWorks, Inc.
%    Copyright 1984-2001
%    The MathWorks, Inc.
%    $Revision: 5.7 $
%    $Date: 2001/04/15 12:01:34 $

t = sum(diag(a));
>> which trace
C:\MATLABR12\toolbox\matlab\...
   matfun\trace.m
```

The *arrow keys* on the computer keyboard can be used to edit the command line. These key arrows will prove very useful when correcting minor misspellings in long input lines. By pressing the up-arrow you can search through the commands you have written previously. By typing the first character, or characters, of a specific command before pressing the up-arrow, only those commands matching this initial input will be scanned through. In R12, a completely new desktop look was presented, where one area of the MATLAB window contains a list of all previous commands. Here you can select and edit any previous call.

The command line is cleared by pressing the escape key (twice on Unix stations). The left and right arrows move the cursor to facilitate deletion and insertion of characters. The control key together with the left and right arrows move the cursor one word forward and backward, respectively, on the current command line. If you are familiar with the text editor *Emacs*, you will find that most of the usual control commands for editing can be used in lieu of the arrow keys. However, some keyboard functions are system dependent, so be aware of that **control-y** on Unix stations may *kill* MATLAB, rather than pasting the buffer!

The table below summarizes the various keys for editing the command line.

↑	Ctrl-p	Recall previous input line
↓	Ctrl-n	Recall next input line
foo ↑ and ↓	Ctrl-p/n	Recall input lines starting with foo
→	Ctrl-f	Step forward one character
←	Ctrl-b	Step backward one character
Ctrl-→	Ctrl-r	Move to the start of the following word to the right
Ctrl-←	Ctrl-l	Move to the start of the previous word to the left
home	Ctrl-a	Move to the beginning of the line
end	Ctrl-e	Move to the end of the line
esc	Ctrl-u	Clear the current line
del	Ctrl-d	Delete the character at the cursor
	Ctrl-k	Delete the rest of the line

Try the commands listed above to learn how to edit the command line.

2   Interactive computation and elementary functions

Content: Interactive calculations, calculation accuracy, elementary functions.
Functions:

General:	`;`, `=`, `:`,
Operators:	`+`, `-`, `*`, `/`
Elementary functions:	`sin`, `cos`, `tan`, `asin`, `acos`, `atan`,
	`sqrt`, `exp`, `log`, `log10`, `round`
Pre-defined constants:	`pi`, `eps`, `realmin`, `realmax`, `Inf`, `NaN`
Complex numbers:	`i`, `j`, `real`, `imag`, `abs`, `angle`, `conj`
Workspace setup:	`format`

A guided tour 4 (Calculator)

To use MATLAB as a calculator, just type the expression you want to evaluate:

$$1 + 2 \cdot 3 - 4^5/6.$$

```
>> 1+2*3-4^5/6

ans =

   -163.6667
```

Note that MATLAB by default displays four decimals in the result of the calculations. This must not be confused with the accuracy that is used to *compute* the result. All computations in MATLAB are performed in *double precision* arithmetic, this means that there are approximately 16 significant figures in the calculations performed in MATLAB.

A guided tour 5 (Controlling output format)

The command **format** controls how the result of computations is displayed in MATLAB. The command **format compact** suppresses the extra empty lines in the output. The default is **format short** which shows at most four decimals, with long format you can view all 16 figures.

```
>> format compact
>> format long
>> 2*3^5/7
ans =
  69.42857142857143
```

The result can also be rounded to a fixed number of decimals with the function **round**. Note that the variable **ans** always contains the numerical result of the last operation in MATLAB. In the help text provided in **round** you find information regarding the related commands **floor**, **ceil** and **fix**. Try them in lieu of **round** in the example to the right.

```
>> format short
>> round(ans*1000)/1000
ans =
  69.4390
>> help round
ROUND Round towards nearest integer.
   ROUND(X) rounds the elements of X
   to the nearest integers.

See also FLOOR, CEIL, FIX.
```

All floating point numbers in MATLAB are stored with a relative precision of roughly 16 decimal digits and they are confined to a finite absolute range of approximately 10^{-308} to 10^{308}. You can use the *pre-defined constants* **eps**, **realmax**, and **realmin** to check the actual values on your computer system. The constant **Inf** is used to indicate a number outside the absolute range, while the constant NaN is an abbreviation for *not-a-number*, indicating a mathematically undefined value. It is important to bare this finite precision in mind when working with MATLAB. The relative accuracy implies that all calculations performed in MATLAB have a round-off error of the order given by the variable eps. Working with numbers of very large or very small magnitude may yield overflow or underflow and perhaps undesirably resulting in an infinite result denoted Inf or an identically zero result as the outcome of some calculations. A detailed background regarding finite precision floating point systems in general, and its effect on the accuracy in calculations in MATLAB in particular, is provided in Section 20.

A guided tour 6 (Elementary functions)

Elementary functions, like $\sin(x)$, $\cos(x)$, $\tan(x)$, $\sqrt{x}$, e^x are evaluated by the functions sin, cos, tan, sqrt, exp respectively. For instance

$$\sin(\pi/4)$$

Note that the constant π is predefined in MATLAB.

An example of another elementary functions is

$$e^{10}$$

note that the letter e in the answer is used to indicate a power-of-ten scale factor.

You can also give the input in the e-notation.

The square root function is also available in MATLAB. Note that everything written behind the sign % is treated as a comment in MATLAB. A complete list of all elementary functions in MATLAB is given by typing help elfun.

The natural logarithm $\ln(x)$ is denoted log in MATLAB, while log10 and log2 denote the base 10 and base 2 logarithms.

You can assign values to symbols and then evaluate an equation:

$$x^2 - x - 1 \qquad \text{for } x = 3$$

Note: A semi-colon ; at the end of a line suppresses the printout.

```
>> sin(pi/4)
ans =
    0.7071
```

```
>> exp(10)
ans =
    2.2026e+04
```

```
>> log(2.2026e4)
ans =
   10.0000
>> sqrt(2) % my comment
ans =
    1.4142
```

```
>> log(exp(10))
ans =
    10
>> log2(65536)
ans =
    16
```

```
>> x=3;   % no printout
>> x^2-x-1
ans =
    5
```

A guided tour 7 (Complex numbers)

Complex numbers are entered using the imaginary unit i or j. They are predefined default as $\sqrt{-1}$ when MATLAB is started, but we can of course define them again.

```
>> i=sqrt(-1)
i =
      0 + 1.0000i
>> z=1+i;
```

The standard operators and functions of real numbers apply to complex numbers as well.

```
>> i*z*sqrt(z)
ans =
     -1.5538 + 0.6436i
```

There are a number of special functions for complex numbers. To compute the absolute value, angle, real and imaginary parts and conjugate, use **abs, angle, real, imag, conj**.

```
>> abs(z)
ans =
      1.4142
```

Exercise 1
Use `format long` to find all decimals in *pi* and *e*, respectively, that are used internally in MATLAB.

Exercise 2
Use `format rat` to find a rational approximation to *pi* and *e*.

Exercise 3
What is largest, e^π or π^e?

Exercise 4
The roots of the equation $x^2 + ax + b = 0$ is given by $x = -\frac{a}{2} \pm \sqrt{\left(\frac{a}{2}\right)^2 - b}$. Use this to solve $x^2 + 3x + 2 = 0$ and verify the solution.

Exercise 5
Verify Euler's equation

$$e^{i\pi} + 1 = 0. \tag{2.1}$$

This is a pretty equation that combines the most important symbols in mathematics in one equation.

Exercise 6
Verify the formula

$$e^{i\phi} = \cos(\phi) + i\sin(\phi) \tag{2.2}$$

for $\phi = 0$, $\pi/4$, π, respectively.

Exercise 7
Compute i^i, where i is the imaginary unit. Despite the simplicity of this equation, it is not easy to solve analytically. We need both a variant of Euler's equation (2.1) and equation (2.2):

$$\log(i^i) = i\log(i) = e^{i\pi/2}i\pi/2 = (\cos(\pi/2) + i\sin(\pi/2))i\pi/2 = -\pi/2$$

and thus $i^i = e^{-\pi/2}$. Verify this result!

3 Manipulating matrices

> **Content:** Entering matrices, operations on matrices.
> **Functions:**
> Operators: `+, -, *, ', .', .*, .^, ./`
> Building matrices: `ones, zeros, eye, diag, rand, randn`
> Elementary matrix functions: `size, length, min, max, sum, sort`
>   `find, prod, cumsum, diff, cumprod,`
>   `reshape`

Matrices constitute the foundational *data type* in MATLAB. In this section we illustrate how to apply different operations on matrices.

A guided tour 8 (Matrices)

Matrices are entered by separating elements either by a space or a comma and rows are separated by a semi-colon.

$$A = \begin{pmatrix} 8 & 1 & 6 \\ 3 & 5 & 7 \\ 4 & 9 & 2 \end{pmatrix}$$

```
>> A=[8 1 6;3,5,7; 4 9 2]
A =
     8     1     6
     3     5     7
     4     9     2
```

A certain element is addressed in an obvious way. The first index is the row number and the second one the column number. To the right, the third element on the first row is extracted.

```
>> A(1,3)
ans =
     6
```

The colon operator can be used to pick out a certain row or column, think of : as "all". Indexing a matrix with only one colon `A(:)` yields a long column vector consisting of all the columns of A stacked on top of each other. Try it to see how it works.

```
>> A(1,:)
ans =
     8     1     6
```

The colon operator will prove very useful and understanding how it works is the key to efficient and convenient usage of MATLAB. The colon operator can also be used to extract a sub-matrix, from A.

```
>> A(:,2:3)
ans =
     1     6
     5     7
     9     2
```

One can address certain arbitrary submatrices using any vectors indexing the columns and rows. Carefully study the example to the right, forming the matrix

$$\begin{pmatrix} A_{22} & A_{21} \\ A_{32} & A_{31} \end{pmatrix}$$

```
>> A([2 3],[2 1])
ans =
     5     3
     9     4
```

The special functions **zeros** and **ones** are used to generate the common matrices of all zero and unit entries. These commands are particularly convenient when forming large matrices where many entries are equal.

Figure out how you would enter an all-zero row vector of length 10, using **zeros**.

Equivalently, we can define b as done to the right.

Unit matrices can be formed with the command **eye**, while general diagonal matrices are created with **diag**, using the diagonal elements provided in a vector argument to this command. Note that **diag** has double purposes. **diag(diag(v))** gives back the vector **v** as the answer and thus it is its own inverse, but **diag(diag(A))** gives back a diagonal matrix from **A**.

The three elementary operations of addition, subtraction and multiplication apply to matrices whenever the dimensions are compatible.

Since MATLAB can handle complex calculus, matrices may have complex entries. Note that the product between a matrix and a scalar can be performed in any order.

Conjugate transpose is entered as a prime ', while non-conjugate transpose of a complex matrix is computed by the operator . '.

The dimensions must match in order to compute the matrix multiplication. The computation to the right shows that all of the sums along the rows of the matrix A are 15.

```
>> b=ones(3,1)
b =
     1
     1
     1


>> b=[1;1;1];

>> C=diag([3 4])
C =
     3     0
     0     4
>> diag(C)
ans =
     3     4


>> C+eye(2)*C-zeros(2)
ans =
     6     0
     0     8


>> J = eye(2) + i*ones(2)
J =
   1.0000+1.0000i        0+1.0000i
        0+1.0000i   1.0000+1.0000i
>> J'*J
ans =
     3     2
     2     3
>> J.'*J
ans =
  -1.0000+2.0000i  -2.0000+2.0000i
  -2.0000+2.0000i  -1.0000+2.0000i
>> A*b
ans =
    15
    15
    15
```

The conjugation has no effect in the case of real matrices so the prime operator can be applied to transpose real matrices. Apparently, the column sums of A are also 15. Check the command **magic** for more information about the matrix A. What happens if you try to enter **b*A** instead?

```
>> b'*A
ans =
      15      15      15
```

You can check the number of rows and columns of any matrix in MATLAB with the command **size**. The vector b has three rows and one column. There is also a command **length** which returns the maximum of these two dimension parameters.

```
>> size(b)
ans =
       3       1
```

Instead of regular matrix multiplication, one can compute the entry-wise multiplication between two matrices of the same dimension using the operator **.***. Compare the results of the two calculations performed to the right.

```
>> A*A
ans =
      91      67      67
      67      91      67
      67      67      91
```

Table 27.1 in Appendix C summarizes the most common matrix functions with one entry-wise and one matrix interpretation.

```
>> A.*A
ans =
      64       1      36
       9      25      49
      16      81       4
```

There are also special operators for the matrix powers illustrated above. A matrix power can be formed with the operator ^, which naturally only is applicable to square matrices. Entry-wise, or array, power is computed with the operator .^ and can be used on matrices of any dimension.

```
>> A.^2
ans =
      64       1      36
       9      25      49
      16      81       4
```

The command **randn** produces matrices with independent random entries from a normal distribution with zero mean and unit variance. How would you produce a Gaussian distributed variable with different mean value and variance? There exists a corresponding function **rand** that produces uniform random variables between 0 and 1.

```
>> D = randn(1,3)
ans =
     -0.9445     -0.4226     -0.3811
```

Addition and subtraction already operate element-wise on matrices and therefore there exists no array version of these operations. However, there is a corresponding array division ./, computing the fraction between each element of two matrices of equal dimension.

```
>> D./D
ans =
       1       1       1
```

The colon operator : can be used for creating a vector of equally spaced values, for example a time scale. The MATLAB syntax 0:0.1:10 yields a vector that starts at 0, takes steps of 0.1 and stops when 10 is reached. There are therefore 101 elements in this vector. Note also that the colon operator always generates row vectors.

```
>> t=0:.1:10;
>> size(t)
ans =
        1    101
```

The **find** function finds the positions where the argument is true. For instance, it is the 51st element of t that is equal to 5. Note that == means a logical comparison, not an assignment.

```
>> ind=find(t==5)
ind =
       51
```

The logical and relational operators available in MATLAB are summarized in the table below, see also **help relop**.

<	less than	&	logical and
>	greater than	|	logical or
<=	less than or equal	~	logical not
>=	greater than or equal	**xor**	exclusive or
==	equal	1	true
~=	not equal	0	false

```
>> find((A(:,2)>b)&(A(:,1)<=3*b))
ans =
        2
>> xor([0 1 0 1],[0 0 1 1])
ans =
        0    1    1    0
```

True and false are represented by the integers 1 and 0. The output from a relative operation is stored in MATLAB as a logical array containing only 1 and 0. Note that the logical comparison operates entry-wise.

```
>> A >= 5
ans =
        1    0    1
        0    1    1
        0    1    0
```

The function **all** returns true if all logical expressions in its vector argument are true. There is also a corresponding function **any**. Note how the matrix A is converted to a column vector using the colon operator.

```
>> all(A(:)>0)
ans =
        1
```

If the argument of an elementary function is a matrix, the output is a matrix of the same dimension, where the elementary function is applied to each element of the matrix argument.

```
>> y=sin(t);
```

The maximum of a vector and its location is computed by max. Similarly, the minimum can be found with min.

Note the way the max function delivers one or multiple output arguments. The function may be called with only one output argument if the index to the maximum is of no importance. This principle holds for all MATLAB functions, they may be called with less than their maximum number of output variables. Check the online help text for the functions to learn how they react on different number of output variables.

```
>> [ymax,ind]=max(y)
ymax =
     0.9996
ind =
    17

>> t(ind)
ans =
     1.6000
```

A vector can be sorted in ascending order using the function sort. This function can also be called with two output arguments. The first output is the sorted input vector, and the second output contains a vector of indices that sort the input argument. Carefully study the example to the right.

Note that MATLAB is case sensitive, the variables a and A are different.

The sum of the elements in a vector is computed by the function sum. Functions related to sum are prod, cumsum, cumprod, and diff.

```
>> a=[1 5 3 9 8];
>> [asort, ind]=sort(a)
asort =
     1     3     5     8     9
ind =
     1     3     2     5     4
>> a(ind)
ans =
     1     3     5     8     9
>> sum(a)
ans =
    26
```

We have exemplified several elementary matrix and vector commands in this section. Type help elmat for a complete listing of the available elementary matrix functions in MATLAB.

The commands sum, sort, max, min etc. have been exemplified on vectors but they can also be applied to matrices. Most of these functions work on each column when applied to a general $(n \times m)$-matrix, find out for yourself by testing them on the matrix A above. Use the help-facilities if you cannot interpret the result directly. The relational operators listed in the tabular above apply entry-wise to matrices just like elementary functions.

Exercise 8

Construct the matrices

$$A = \begin{pmatrix} 1 & 2 & \cdots & 10 \\ 2 & 2 & 0 & 0 \\ \vdots & 0 & \ddots & 0 \\ 10 & 0 & 0 & 10 \end{pmatrix} \qquad M = \begin{pmatrix} 2 & 0 & 0 & 0 & 0 \\ 0 & 2 & 0 & 0 & 0 \\ 3 & 3 & 3 & 4 & 0 \\ 3 & 3 & 3 & 0 & 4 \end{pmatrix}$$

using the matlab functions diag, :, ones, zeros and eye. Try the commands pcolor and spy on the two matrices.

Exercise 9
Construct the matrix

$$A = \begin{pmatrix} 1 & 2 & \dots & m \\ m+1 & m+2 & \dots & 2m \\ \vdots & & & \vdots \\ (n-1)m+1 & (n-1)m+2 & \dots & nm \end{pmatrix}$$

using the reshape *function. Test it for the case* $m = 7$ *and* $n = 8$.

Exercise 10
A matrix containing a magic square of size n *is generated with* magic. *Type* A=magic(n); *and verify that the sum of each row and each column is the same. Also verify that this constant is* $s = (n^3 + n)/2$.

Exercise 11
Verify that v=ones(n,1) *is an eigenvector of* A *in Exercise 10 with eigenvalue* s, *using the definition of eigenvalue and eigenvector* $Av = sv$.

Exercise 12
Let P=2*magic(n)/(n^3+n) *such that the row and column sum is one. Such a matrix has a special meaning in statistics, since the element* A(i,j) *can represent the probability to go from state* i *to state* j *in a Markov chain. Verify that* P^n *converges to the matrix* 1/n*ones(n).

Exercise 13
Compute a list of factorials $(1!, 2!, 3!, \dots, n!)$ *using* cumprod.

Exercise 14
Generate a Gaussian random vector with mean μ *and covariance matrix* P *given by*

$$\mu = \begin{pmatrix} 1 \\ 2 \end{pmatrix} \qquad P = \begin{pmatrix} 20 & 16 \\ 16 & 20 \end{pmatrix} = \begin{pmatrix} 4 & 2 \\ 2 & 4 \end{pmatrix}^2$$

using the transformation

$$x = \begin{pmatrix} 1 \\ 2 \end{pmatrix} + \begin{pmatrix} 4 & 2 \\ 2 & 4 \end{pmatrix} x_o$$

where x_0 *is normalized and can be generated by* randn.

Exercise 15
Compute the scalar product between the vectors $(1, 2, \dots, 100)$ *and* $(1, 1, \dots, 1)$, *both of dimension* 1×100.

Exercise 16
Let $S_n = \sum_{k=1}^{n} \frac{6}{k^2}$. *Compute* $\sqrt{S_n}$ *for increasing values of* n, *do not stop until you reach* $n \geq 100000$. *Can you guess what the theoretical limit is?*

Exercise 17
Generate 100 random integers uniformly distributed from 0 and 1000. Find the maximum rmax *and its position* imax, *verify that* r(imax)==rmax. *Sort the integers in ascending order, verify that* rsort(1)==min(r) *and that* rsort(100)==max(r). *Compute the number of integers larger than 900. Compute the sum of all the integers.*

Exercise 18

Find a numerical approximation to

$$\int_0^1 \frac{1}{\sqrt{t^2 + 2}} \, dt$$

and compare it to the analytical solution. Hint: Construct a grid over the range of t, dense enough to make the integrand approximately constant over each interval, and use the function sum.

Exercise 19

Generate a sinusoid of one period. Find a numerical approximation of

$$\frac{1}{2\pi} \int_0^{2\pi} I(|\sin(x)| > 0.9) \, dx.$$

Here I is the indicator function, which is one whenever the argument is true and zero otherwise. Hint: Use the abs and the find functions.

4   Strings and workspace administration

Content: Entering and manipulating strings and cell arrays.
Functions:
Construction and conversion: '', {}, char, str2num, num2str, double,
                                str2mat
Operations: clear, disp, eval, strcmp, upper, lower

A guided tour 9 (Character strings)

A *string* is entered by using single quotes '.
A quote within the string is indicated by two
quotes, *e.g.* **why='Don''t ask!'** (The response to **why(6)** in MATLAB).

```
>> s='abcdefgh'
s =
abcdefgh
```

A string is treated and stored as any vector
in MATLAB. Internally, the vector contains
numerical values that code each character and
the length of the vector is therefore equal to
the number of characters in the string. The
actual printout depends on the character set
encoding for the given font.

```
>> p='ijklmn';
>> q=[s p 'opqr']
q =
abcdefghijklmnopqr
```

```
>> q(13)
ans =
m
```

A character string can be converted to numbers using the function **double**. The conversion is reversed by the command **char**.

```
>> pnum = double(p)
pnum =
   105   106   107   108   109   110
>> pnew = char(pnum)
pnew =
ijklmn
```

The printable characters in the basic ASCII
character set are represented by the values 32
to 127, while the values 128 to 255 represent
the extended character set and the characters
encoded from these values will depend on the
character set of the computer.

```
>> s='10';
>> 2*s
ans =
    98    96
```

Can you figure out how **2*s** is interpreted by
MATLAB? Check the ASCII encoding by typing **char([32:79;80:127])**

```
>> 2*str2num(s)
ans =
    20
```

A very common operation is to mix fix strings
with numerical results and string variables.
The example illustrates the use of **num2str**
to convert a real number to a string, and the
function **disp** to display the result.

```
>> units='pixels';
>> size=12;
>> disp(['The size is ',...
      num2str(size),' ',units])
The size is 12 pixels
```

Strings that are mixtures of characters and numerical values can also be evaluated. In this example, the numerical value might be given by some preceding calculations, and we want to compute the sinusoid. The function **eval** evaluates strings, and will be further described in guided tour 25.

Complete strings can be tested for equality using the command **strcmp**. The functions **upper** and **lower** are convenient when one wants to test string equality independent on if the string is upper or lower case.

To create larger text material, as the automated letter to the right, the function **str2mat** is useful. It constructs a matrix of strings, where each string is padded with spaces so all strings get the same number of elements. The example uses the Unix **mailx** command to send mails automatically, and the *exclamation mark !* is the prefix for *system calls*.

As an alternative to **str2mat**, a *comma separated list* of strings can be used:

```
>> cellmessage={['Dear ',name],...
   negative, closing}
```

Here each string is one element in the *cell array* **cellmessage**. Conversion between a cell array and matrix padded with spaces is done with

```
message=char(cellmessage)
cellmessage=cellstr(message)
```

```
>> x=5;
>> command=['sin(',num2str(x),')'];
>> disp(command)
sin(5)
>> eval(command)
ans =
    -0.9589
>> strcmp(p,pnew)
ans =
     1
>> strcmp(lower('IjKlMN'),p)
ans =
     1
>> negative=str2mat(...
'We regret to inform you,',...
'your application has been rejected');
>> closing='Sincerely, Big Boss';

>> name='Mr. Looser';
>> email='looser@hotmail.com';
>> message=str2mat(['Dear ',name],...
negative,closing)
message =
Dear Mr. Looser
We regret to inform you,
your application has been rejected
Sincerely, Big Boss
>> !mailx email message
```

The guided tour above defines a brief introduction to how strings are handled in MATLAB. Try typing **help strfun** for a complete list of various commands for operating on character strings.

A guided tour 10 (Administrating the workspace)

The function **who** gives information about the variables available in the workspace.

```
>> who

Your variables are:

A           D           ans         p
C           J           b           s
```

The command **whos** gives full information
about the size and type of each variable defined in the MATLAB workspace. Also try typing **workspace** for an interactive, menu driven,
workspace browser (available from R10).

```
>> workspace
```

```
>> whos
   Name        Size            Bytes  Class

   A           3x3                72  double array
   C           2x2                32  double array
   D           1x3                24  double array
   J           2x2                64  double array (complex)
   ans         1x1                 8  double array (logical)
   b           3x1                24  double array
   p           1x6                12  char array
   s           5x3                30  char array

Grand total is 24 elements using 232 bytes
```

The variables in the workspace can all be
stored on a file. To save all of them on the file
mydata.mat, follow the example to the right.
Note that the file extension **.mat** is added automatically. The second example shows how
to save only the variables *A*, *b* and *C* in the
file **myABC.mat**.

```
>> save mydata
>> save myABC A b C
```

Data can also be saved and loaded in plain
text ASCII format using the switch **-ascii**

By clearing the workspace all the current variables are removed, but we can reload them all
back again from the file **mydata**.

```
>> clear
>> who
>> load mydata
```

The function **diary** is useful if you want to
save a complete MATLAB session. It saves all
function calls and answers as they appear in
the MATLAB window on a text file.

```
>> diary mysession
>> sin(2)
ans =
    0.9093
>> diary off
```

You can use the MATLAB function **type** to
view the diary file. The diary file is a standard
text file which you can edit in a text editor or
print on a printer.

```
>> type mysession

>> sin(2)
ans =
    0.9093
>> diary off
```

The files `mydata.mat`, `myABC.mat`, and `mysession` are stored in the current working directory. You can list all the files in this directory with the command `dir`. The command `pwd` is an acronym for print working directory. The working directory and the search path in MATLAB are discussed in detail in Section 8.

Syntactically, there is basically no difference between a function and a matrix. In an expression `y=fun(x)`, MATLAB first looks for a variable with the name `fun` and then, if no such variable exist, a function called `fun`. Note that it is possible to redefine functions as variables, as seen in the example to the right.

It is therefore advisory to use unique variable names that do not conflict with any MATLAB functions or predefined variables. The matrix can be deleted from the workspace using `clear` and specifying what variable should be removed. The function definition of `sin` is then valid once again.

```
>> dir
.                     ..
myABC.mat   mydata.mat mysession
>> pwd
ans =
C:\matlabr12\work
>> sin=1:5;
>> sin(2)
ans =
     2


>> clear sin
>> sin(2)
ans =
       0.9093
```

Exercise 20
Define a matrix that produces the following printout:

```
>> A
A =
     one
     two
   three
```

Exercise 21
Compute an ASCII table of all elements between 32 and 127, where a part of the table looks as follows:

```
 97  a
 98  b
 99  c
100  d
```

Exercise 22
Generate a random text consisting of 17000 upper case letters A to Z. How many times does the letter A occur? What is the expected number of occurrences of one of 26 characters in 17000 random samples? Hint: the ASCII code for A is 65 and for Z is 90.

Exercise 23
Generate a 26 × 3 matrix containing a list of random TLA's (three letter acronyms), as the small example shown below

```
>> s
s =
```

```
BCE
UCJ
DGJ
PLS
ORD
```

Order the generated list of your acronyms alphabetically by the first letter.

Exercise 24
Consider the following string scrambler:

```
str = 'The spy must die';
tmp = reshape(str,[4 4])'
tmp =
The
spy
must
 die
codestr = tmp(:)'
codestr =
Tsm hpudeysi  te
```

Construct an unscrambler that recovers the message.

Exercise 25
Consider the following encryption function:

```
>> s='THESPYMUSTDIE';
>> scoded=char(64+mod(diff([0 double(s)-64]),26));
```

Figure out how it works, and construct a decryption function! Hint: compute and examine the matrix

```
>> [double(s)-64;diff([0 double(s)-64]);mod(diff([0 double(s)-64]),26)];
```

Exercise 26
The idea of prime number codes is to find a large number which is a product of two prime numbers. The factorization can take very long time to compute if the number is large enough. Use the function primes *to generate prime numbers up to* $n \leq 10^5$. *Take the two largest prime numbers less than* n *and compute their product, which then is public key, which can be exposed to anyone:*

```
>> key=prime1*prime2;
```

There is a sophisticated method to encode a string to an illegible message, which can only be decoded by knowledge of the primes.

Use factor *to factorize the product, and confirm that the computation time grows with* n *using* tic, toc. *Since the system clock, which is used by* tic, toc, *often has a resolution of 1 microsecond, repeat each factorization 100 times inside the timing. Examine for instance the values* $n = 100, 200, 500, 1000, 2000, 5000, 10000, 20000, 50000$. *This means, that the computational complexity to break the code increases in the key size. The memory requirement increases even faster, a problem is solved brutally in the* MATLAB *implementation by putting an upper bound of* 2^{32} *on the argument of* factor. *A key with 128 bits (corresponds to 38 decimal digits) is today considered to be unbreakable.*

Exercise 27

A simple but powerful encryption scheme is given below:

```
>> rand('seed',key)
>> s=['The Spy must die';'Please, confirm!'];
>> svec=double(s);
>> e=round(256*rand(size(svec)));
>> sveccoded=mod(svec+e,256);
>> scoded=char(sveccoded)
scoded =
d4T7\Ǔ{16c+h?'u^\gX6dfj~R5)fd
```

Construct the decryption scheme!

5   Graphical illustrations

Content: Plotting facilities.
Functions:
plot, hist, stem, semilogy, semilogx, loglog,
title, xlabel, ylabel, text, gtext, grid, axis,
figure, subplot, hold, legend, ginput, zoom, print

A guided tour 11 (Elementary Graphics)
The function plot(x) plots the values in the
vector x in a figure window. The values in x
are plotted on the vertical axis against the vector index number on the horizontal axis. Linear interpolation is used to generate a solid
line graph.   The related functions loglog,
semilogx, semilogy produces simular plots
using logarithmic axes.

Note how the colon operator is used to generate a descending sequence in the definition of
x.

```
>> x=[1:4 5:-1:0 4 2 5];
>> plot(x)
```

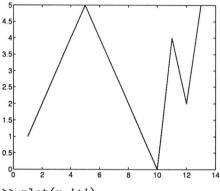

The color and the line type of the graph can
be specified, see **help plot** for details.   Instead of interpolation between the the points
in the vector x, a symbol like * can be used to
indicate discrete data points. Other available
maker styles are listed below.

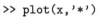

s square	< triangle (left)	+ plus
x x-mark	^ triangle (up)	. point
d diamond	v triangle (down)	* star
o circle	> triangle (right)	
h hexagram	p pentagram	

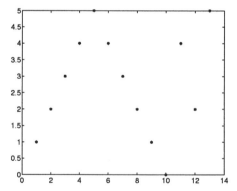

Both an interpolated curve and a star at each point can be obtained using the **hold** function. With **hold on** each new plot is overlayed on the previous ones. The command **hold off** resets the current figure window to its default behavior. Alternatively, the function

```
>> clf
```

can be used, it is an abbreviation for clear current figure which resets the figure window and clears all previous plots from it.

Alternatively, in later versions of MATLAB the figure to the right can be produced by combining the marker and line styles in one command

```
>> plot(x,'-*')
```

The visualization shown to the right is convenient for a discrete series of numbers. Its calling syntax is similar to the one used for the plot function described above. The plot functions **plot**, **stem**, etc. automatically opens a new figure window the first time they are invoked. Subsequent plot commands will use in the same window unless a new figure window is opened by issuing the command **figure**. To switch back to the first figure window, type

```
>> figure(1), gcf
ans =
       1
```

The number of the current figure window is displayed with the command **gcf**, get current figure.

```
>> hold on
>> plot(x)
>> hold off
```

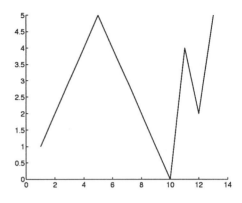

```
>> stem(x)
```

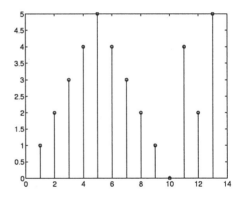

If the function **plot** is called with two vector arguments, the first one corresponds to the labeling of the horizontal axis. Carefully study the example to the right and note the different response to the calls **plot(t,sin(t))** and **plot(sin(t))**.

```
>> t=0:0.01:10;
>> plot(t,sin(t))
```

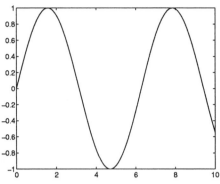

Several plots can be plotted at the same time using the **hold** function as described above, but it is often more convenient to form a matrix of each column vector to be plotted. Note that if plot is called with a matrix argument, each columns is interpreted as a sequence of numbers to plot. Since **y1** and **y2** are row vectors (why?), they need to be transposed.

```
>> y1=sin(t);
>> y2=sin(t.^2);
>> plot(t,[y1' y2'])
```

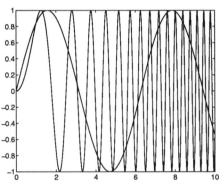

Another possibility to plot several vectors and still enable a free choice of line types, colors etc. is shown to the right. In this example, the first curve is a solid yellow line and the second one is dotted and green. The available color characters are

y yellow	m magenta	w white
c cyan	r red	k black
g green	b blue	

and the linestyles are

–	solid
--	dashed
-.	dash-dotted
:	dotted

```
>> plot(t,sin(t),'y-',...
        t,sin(t.^2),'g:')
```

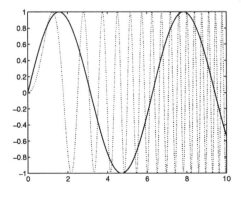

The plot can be modified in several ways as the following example illustrates. See also **gtext**, where you can put text anywhere in the plot window using the mouse. The **grid** function has a calling syntax similar to the **hold** function.

The command

```
>> grid on
```

turns on the grid and the command

```
>> grid off
```

turns it off, issuing the command **grid** by itself, without the either of the keywords **on** or **off**, alternates between the two states of having the grid visible and invisible. There is also a command **zoom** having the same calling syntax. With this command you can use the mouse to zoom in your plots. Experiment with it.

The question posed in the figure above can be answered with the function **legend**. The command **legend off** removes the legend.

Note how **axis** is used above to reset the automatic range of the axes produced by the plot command. This function also accepts some special keywords. The command **axis equal** sets the length of the both axes equal, while **axis square** sets the scale equal on both axes.

```
>> title('Sinusoid and chirp')
>> xlabel('Time')
>> ylabel('Amplitude')
>> axis([0 5 -2 2])
>> grid
>> text(3,1.3,'Which is which?')
```

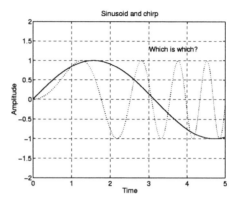

```
>> legend('sin','chirp')
```

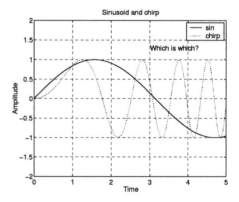

The `subplot` function creates an array of plots in the same figure window. For instance, `subplot(211)`, or `subplot(2,1,1)`, divides a figure window into two rows and one column and makes the first sub-plot active. This is suitable for compact printouts. The command `print` sends the current figure window to the default printer connected to the computer. There is also a **Print** option in the **File** menu on the figure window providing easy access to different printing options.

```
>> subplot(211)
>> plot(t,sin(t))
>> title('Sinusoid')
>> xlabel('Time')
>> ylabel('Amplitude')
>> axis([0 5 -2 2])
>> subplot(212)
>> plot(t,sin(t.^2))
>> title('Chirp')
>> xlabel('Time')
>> ylabel('Amplitude')
>> axis([0 5 -2 2])
>> subplot(111)
```

The command

```
>> print -deps myfile.eps
```

saves the current figure window as an Encapsulated Level 1 *PostScript* file called `myfile.eps`. This file can be included in all word processors that can interpret PostScript objects. On PC systems running Windows there is often no support for PostScript files. Then it is more convenient to choose **Copy Figure** in the **Edit** menu and thereafter paste the figure directly into the word processor. If this option is pursued, make sure that the **Copy Options** is set to **Windows Metafile**.

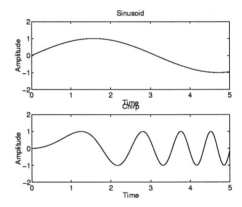

-deps	.eps	Postscript
-depsc	.eps	Postscript with color
-dtiff	.tif	TIFF
-djpeg90	.jpg	JPEG, with quality $90 \in [0, 100]$.
-dpng	.png	Portable Network Graphic
-dill	.ill	*Adobe Illustrator* format
-dmeta	.wmf	*Microsoft* meta format
-dbitmap	.bmp	*Microsoft* bitmap format
-r150		Resolution for ps, tif, jpg and meta in dots/inches

Table 5.1 Print options.

Exercise 28
Illustrate the function

$$f(x) = \frac{1}{(x - 0.3)^2 + 0.01} + \frac{1}{(x - 0.9)^2 + 0.04} - 6$$

over the interesting x-range. Use `ginput` to find the local minimum and maxima (see help documentation for `ginput`). Also try the `zoom` and `grid` commands.

Exercise 29
Plot the parametric function

$$x(t) = \cos\left(-\frac{11t}{4}\right) + 7\cos(t)$$

$$y(t) = \sin\left(-\frac{11t}{4}\right) + 7\sin(t)$$

over the range $0 < t < 8\pi$.

Exercise 30
Plot the function

$$f(t) = t\cos(\omega t) + it\sin(\omega t)$$

in the complex plane. Use $\omega = 1, 2, 3$ and plot the three functions in the same window. What is the effect of ω? A suitable interval for t is from 0 to 10 with steps of 0.001.

Exercise 31
We want to transmit the signal $\sin(\omega_s t)$ on a radio channel using the carrier $\sin(\omega_c t)$. Illustrate the difference in AM and FM by comparing the signals

$$\text{AM}: \quad \sin(\omega_s t)\sin(\omega_c t)$$
$$\text{FM}: \quad \sin(\omega_c t + 5\sin(\omega_s t))$$

as pedagogically as possible. Use for instance $0 < t < 10$, $\omega_c = 10$ and $\omega_s = 1$.

Exercise 32
Produce two plots similar to the following ones:

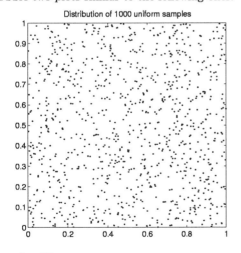
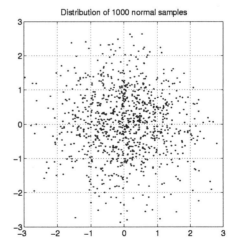

Exercise 33
The factorial function n! grows approximately as

$$(n-1)! \approx f(n) = \sqrt{2\pi}n^{n-1/2}e^{-n}.$$

This is called Stirling's formula. Verify the formula by plotting $(n-1)!/f(n)$. Also plot horizontal lines at 0.99 and 1.01, respectively. For how large n is the approximation better than 1%?

6   Matrix algebra and polynomials

> **Content:** Basic linear algebra and matrix computations.
> **Operators:** \
> **Functions:**
> inv, eig, det, svd, rank, cond, rcond, rref
> sqrtm, expm, conv, roots, poly, polyval, null

See Tables 27.1 and 27.2 in Appendix C for a summary of linear algebra concepts and these MATLAB functions, and Table 27.6 for a summary of polynomials.

A guided tour 12 (Polynomials)

Polynomials are represented as vectors in MATLAB. This example creates the polynomial below and evaluates it at a given point by polyval.

```
>> p1 = [1 2 3 4];
>> polyval(p1,2)
ans =
    26
```

$$p_1(x) = x^3 + 2x^2 + 3x + 4 \qquad p_1(2) = 26$$

Suppose we are interested in the product $p(x) = p_1(x)p_2(x)$ where the polynomial $p_2(x) = x + 1$. The coefficients of $p(x)$ are then found by convolving the two coefficient vectors. The result is

```
>> p2=[1 1];
>> p=conv(p1,p2)
p =
    1    3    5    7    4
```

$$p(x) = (x+1)p_1(x) = x^4 + 3x^3 + 5x^2 + 7x + 4$$

The roots of $p(x)$ are defined as the coefficients r_i in

$$p(x) = (x + r_1)(x + r_2)(x + r_3)(x + r_4)$$

and they are computed by roots. These are of course the union of the roots of $p_1(x)$ and $p_2(x)$. These roots can be illustrated in the complex plane, as shown here. Note that since r is a vector of complex numbers, plot switches automatically to the complex plane and plots the imaginary part of r against its real part. That is, whenever a vector z has an imaginary part

```
>> r=roots(p)
r =
    -0.1747 + 1.5469i
    -0.1747 - 1.5469i
    -1.6506
    -1.0000
>> plot(r,'s')
```

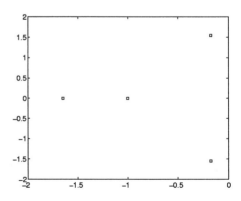

```
>> plot(z)
```

is equivalent to

```
>> plot(real(z),imag(z))
```

A guided tour 13 (Matrix Algebra)

Consider the following matrix.

$$A = \begin{pmatrix} 1 & 2 & 3 \\ 4 & 5 & 6 \\ 7 & 8 & 0 \end{pmatrix}$$

The function **inv** determines the matrix inverse. Analytically,

$$A^{-1} = \frac{1}{9} \begin{pmatrix} -16 & 8 & -1 \\ 14 & -7 & 2 \\ -1 & 2 & -1 \end{pmatrix}$$

The function **det** computes the determinant,

$$\det A = 27$$

The rank of a matrix is defined as the maximal number of linearly independent rows or columns, it is determined by the function **rank**.

A matrix eigenvalue λ is a scalar number such that $Ax = \lambda x$ for some vector x, the corresponding eigenvector to the eigenvalue λ. The eigenvalues of a matrix are computed by the function **eig**.

The eigenvectors are also provided by **eig** if an extra output variable V is defined. Note the subtle difference between the call to **eig** above and to the right. When the function is called with two output arguments the eigenvalues are provided in the second argument as a diagonal matrix. This kind of different behavior depending on the number of output variables and the type of input variables is common in many MATLAB functions. It is advisory to always read the on-line help documentation carefully when calling functions with different types of inputs or different number of outputs than usual.

The singular values of a matrix A are the square roots of the eigenvalues of $A^T A$. The singular values are determined by the function **svd**, which has a calling syntax similar to **eig**. The singular values are sorted in descending order.

```
>> A=[1 2 3; 4 5 6; 7 8 0]
A =
     1     2     3
     4     5     6
     7     8     0
>> inv(A)
ans =
   -1.7778    0.8889   -0.1111
    1.5556   -0.7778    0.2222
   -0.1111    0.2222   -0.1111
>> det(A)
ans =
    27
>> rank(A)
ans =
     3
>> eig(A)
ans =
   12.1229
   -0.3884
   -5.7345
>> [V,D]=eig(A)
V =
    0.7471   -0.2998   -0.2763
   -0.6582   -0.7075   -0.3884
    0.0931   -0.6400    0.8791

D =
   -0.3884         0         0
         0   12.1229         0
         0         0   -5.7345


>> s=svd(A)
s =
   13.2015
    5.4388
    0.3760
```

The matrix norm is defined as the largest singular value. The function **norm** is another example of a function having a polymorphic characteristic. When called with a matrix argument, as illustrated to the right, it computes a matrix norm. When **norm** is called with a vector argument it computes a vector norm. See **help norm** for details.

```
>> norm(A)
ans =
    13.2015
>> norm(A*V-V*D)
ans =
    2.0400e-014
```

Note that the computed eigenvalues and eigenvectors cannot be determined exactly due to round-off errors.

The *condition number* is defined as the ratio between the largest and smallest singular values, and it is computed by the function **cond** in MATLAB. It is commonly used as an indicator on how easy it is to compute its inverse, or to solve a linear system of equations $Ax = b$ (see next example). A matrix having a large condition number indicates that it is close to being singular.

```
>> cond(A)
ans =
    35.1059
```

Consider the linear equation system:

$$\begin{pmatrix} 1 & 2 & 3 \\ 4 & 5 & 6 \\ 7 & 8 & 0 \end{pmatrix} x = \begin{pmatrix} 1 \\ 1 \\ 1 \end{pmatrix}$$

The obvious way to solve for x is to use the matrix inverse.

```
>> b=[1;1;1];
>> x=inv(A)*b
x =
   -1.0000
    1.0000
   -0.0000
```

However, a numerically much more reliable way is to use the backslash operator \. This operator does not invert the matrix explicitly, but uses a factorization approach. Furthermore, the backslash can be used for under-determined and over-determined systems of equations. That is, systems of equations having more unknowns than equations (it then computes the minimum norm solution) and less unknowns than equations (when the *least squares* solution is computed), respectively. See **help slash** for details.

```
>> x=A\b
x =
   -1.0000
    1.0000
   -0.0000
```

Some well-known elementary functions can be extended to matrices. For instance, the matrix square root B_2 of a matrix B is defined such that $B_2 B_2 = B$. It is computed by the MATLAB function B2=sqrtm(B). Note that the call B2=sqrt(B) applies the square root to each element of the matrix argument. Also note that MATLAB, due to numerical round-off errors, does not realize that the matrix elements of $B_2 B_2$ actually are real rather than complex numbers. To the right, the real part can be extracted using the command real(B2*B2).

Another example of a matrix function is the matrix exponential e^A, extensively used in linear systems and control theory. In analogy with the scalar exponential, it is defined such that

$$\frac{d}{dx}e^{Ax} = Ae^{Ax}$$

and computed with expm.

The function poly determines the coefficients of the characteristic polynomial to a matrix. In MATLAB, a polynomial is represented by a vector containing the coefficients in descending powers of the indeterminate, that is

$$p(\lambda) = \det(A - \lambda I) = \lambda^3 - 6\lambda^2 - 72\lambda - 27$$

An interesting property in matrix theory is $p(A) = 0$, which is here verified for this example. (This is called the *Cayley-Hamilton theorem.*)

The roots of the characteristic polynomial are the eigenvalues of the matrix. Calling poly with a vector argument yields a polynomial with roots given by the vector argument. In this setting, the functions poly and roots are inverses of each other up to ordering, scaling, and roundoff error.

```
>> B = A(1:2,1:2);
>> B2=sqrtm(B)
ans =
  0.5373+0.5373i  0.7339-0.1967i
  1.4679-0.3933i  2.0052+0.1440i
>> B2*B2
ans =
  1.0000-0.0000i 2.0000
  4.0000-0.0000i 5.0000-0.0000i
>> norm(B-B2*B2)
ans =
  1.0801e-015
```

```
>> expm(A)
ans =
  1.0e+04 *
    3.1591      3.9741      2.7487
    7.4540      9.3775      6.4858
    6.7431      8.4830      5.8672
```

```
>> p=poly(A)
p =
  1.00  -6.00  -72.00  -27.00
>> A^3 - 6*A^2 - 72*A - 27*eye(3)
ans =
     0      0      0
     0      0      0
     0      0      0
```

```
>> r=roots(p)
r =
   12.1229
   -5.7345
   -0.3884
>> poly(r)
ans =
     1.00  -6.00  -72.00  -27.00
```

There are many matrices giving the same characteristic polynomial and thus the same eigenvalues. One example of how to construct a matrix with specified characteristic polynomial is computed with compan, which computes the *companion matrix*.

```
>> Acomp=compan(p)
Acomp =
    6.0000   72.0000   27.0000
    1.0000          0          0
         0    1.0000          0
>> p=poly(Acomp)
p =
    1.00   -6.00   -72.00   -27.00
>> eig(Acomp)
ans =
   12.1229
   -5.7345
   -0.3884
```

Above, several elementary matrix functions and operators are exemplified. The available functionality in MATLAB is far greater, see help matfun and help polyfun for comprehensive listings of matrix and polynomial functions in MATLAB.

Exercise 34
Verify the useful formulas $\det(I+AB) = \det(I+BA)$ and $\text{trace}(AB) = \text{trace}(BA)$ for two randomly generated matrices A and B of size (3×3).

Exercise 35
Verify the matrix inversion lemma

$$(A + BCD)^{-1} = A^{-1} - A^{-1}B(C^{-1} + DA^{-1}B)^{-1}DA^{-1}$$

numerically by using A=magic(3) and

$$B = \begin{pmatrix} 1 & 0 & 0 \end{pmatrix}^T, \quad C = 2, \quad D = \begin{pmatrix} 1 & 0 & 0 \end{pmatrix}.$$

Exercise 36
Verify the block matrix inversion formula

$$\begin{pmatrix} A & B \\ D & C \end{pmatrix}^{-1} = \begin{pmatrix} A^{-1} + A^{-1}B\Delta^{-1}DA^{-1} & -A^{-1}B\Delta^{-1} \\ -\Delta^{-1}DA^{-1} & \Delta^{-1} \end{pmatrix}$$

where $\Delta = C - DA^{-1}B$ numerically testing the matrices A, B, C, D in Exercise 35.

Exercise 37
The functions poly and root are inverses of each other. Define the following two vectors in MATLAB and evaluate the relational expressions. Explain the result.

```
>> r = [1; 2]; p = poly(r);
>> r == roots(poly(r)), p == poly(roots(p))
```

Exercise 38
Set r=[1 2 3 4 5 6] and plot the corresponding polynomial $p(x)$ for $0 \leq x \leq 7$. Differentiate the polynomial by hand, compute the local maxima and minima and compare with the plot.

7 Advanced graphics

> **Content:** Illustrations in three dimensions.
> **Functions:**
> contour, mesh, meshgrid, view, rotate3d, surf, surfc, surfl,
> colormap, contour3, clabel, colorbar, propedit

There are basically two different ways of illustrating a three-dimensional function: mesh and contour. The first one gives a three-dimensional plot while the second looks at the surface from above and level curves and colors illustrate the height. There are several variants of these, and the color map, shading, viewpoint and so on can be changed.

A guided tour 14 (Three dimensional graphics)

Consider the function

$$f(x, y) = (x^2 + 3y^2)e^{1-x^2-y^2}$$

in the area defined by $|x| < 2$ and $|y| <$ 2.5. First, construct two one-dimensional grids along the x and y axes with a step-size of 0.1. Note that these two vectors have different length.

```
>> x=-2:0.1:2;
>> y=-2.5:.1:2.5;
>> size(x),size(y)
ans =
     1    41
ans =
     1    51
```

A two-dimensional grid is given by two matrices X and Y consisting of repeated copies of the vectors x and y along their respective rows and columns. The values of the function $f(x, y)$ in the sought area is found be evaluating the function entry-wise over these two matrices. The three matrices X, Y, and f therefore have equal dimension determined by the length of the vectors x and y, respectively.

```
>> [X,Y]=meshgrid(x,y);
>> f = (X.^2+3*Y.^2).*...
       exp(1-X.^2-Y.^2);
>> size(f)
ans =
    51    41
```

The function mesh illustrates the matrix f and gives a three dimensional view of the function. Plotting commands related to mesh are surf, meshc, meshz, and waterfall. Use surfl or surfc in lieu of mesh in the example to the right. Try a different color map, for example

```
>> mesh(x,y,f);
```

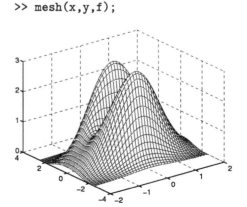

>> surfl(f); colormap bone; colorbar

Experiment with the different color maps listed when typing **help graph3d**. The command **colormap default** returns to the default colors. The color is only one of several adjustable properties of the figure window. The figure shading, lighting and material properties, and viewpoint are also at the user's hand.

For example, the default viewpoint of -37.5 degrees azimuth and 30 degrees elevation can be changed to -60 and 50 degrees, respectively. You can also set the viewpoint manually, using the mouse pointer. Issue the command

```
>> rotate3d
```

press the left mouse button and drag in the current figure window. The plot is re-rendered when you release the mouse button. The command rotate3d has a calling syntax similar to zoom. rotate3d and zoom are also available as icons on the figure window (introduced in R11) and can be adjusted along with several other properties of the figure, see Editing Plots under the Help menu of the figure.

The contour plot gives a two dimensional (x, y) plot, where the constant level curves $f(x, y) = \kappa$ are plotted for a number of different constants chosen automatically. One can also force the function to show a fixed number of contours, and the level curves can be labeled with the constant values they illustrate. Try for example

```
>> C=contour(x,y,f,6);
>> clabel(C)
```

The matrix C is a contour matrix that defines the level curves, see help contourc for details.

The values of the level curves can also be defined by the user. For instance, only the x, y values corresponding to $f(x, y) = 1$ and $f(x, y) = 2$ can be illustrated. There is also a related function contour3 which displays the contours on their true vertical level in a three-dimensional plot.

The graphics in MATLAB are object oriented. This means that everything that appears in the plots are objects whose properties can be changed. This will be described in more detail in Section 13.

```
>> axis([-2 2 -2.5 2.5 0 3])
>> view(-60,50)
```

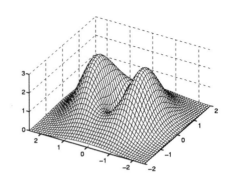

```
>> contour(x,y,f);
```

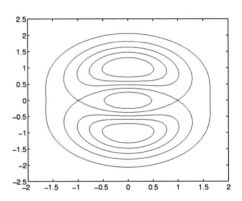

```
>> contour(x,y,f,[1 2]);
```

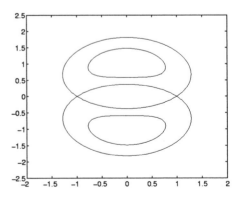

The function **propedit** is a graphical user interface for accessing and changing the properties of all figures. Here you can change colors, font shapes and sizes, line colors and widths *etc.*. This is one alternative to get graphics suitable for publications, see also Section 27.

The current state of the figure along with all the properties that have been altered can be saved to a file for later reference and further adjustments (introduced in R11). Just choose **Save** in the **File** menu of the figure to save it with extension `.fig`. The figure can be reloaded from the same menu, for example during another MATLAB session.

```
>> propedit
```

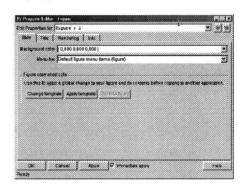

Further details and lists of visualization commands are given by issuing `help graph3d` and `help specgraph`.

Exercise 39
Use `mesh`, `view`, and `rotate3d` to examine the function

$$f(x,y) = 3xe^y - x^3 - e^{3y}, \quad -\infty < x < \infty, \ -\infty < y < \infty$$

The tangent function can be used to compress the entire plane to a finite square. Therefore, plot instead
$$\tan^{-1}(f(\tan(s), \tan(t))), \quad -\pi/2 < s < \pi/2, \ -\pi/2 < t < \pi/2$$
This function has only one critical point, but the local maximum is not an absolute maximum, which occurs on the boundary $y \to \infty$. This cannot happen to a function in one variable, which must have another critical point, namely a local minimum, for this to happen.

Exercise 40
Illustrate the function

$$f(x,y) = \begin{cases} \frac{xy(x^2-y^2)}{x^2+y^2} & \text{if } (x,y) \neq 0 \\ 0 & \text{if } (x,y) = 0 \end{cases}$$

The function is continuous at the origin with continuous derivatives. This is a well-known example where the mixed partial derivatives are not equal $f_{xy}(0,0) \neq f_{yx}(0,0)$, but this is perhaps not evident from graphics.

Exercise 41
Issue the command `logo` which generates a figure window with the MATLAB logo. View the logo m-file and experiment with the visualization commands in MATLAB. Try for example

```
>> cameramenu
```

which gives an extra menu in the figure window where you can alter the camera and lighting properties of the figure. Experiment with the settings!

8   MATLAB **Scripts**

Content: Script files and the MATLAB programming environment.
Functions:
`cd, dir, !, mkdir, path, addpath, rmpath, pathtool, edit, which,`
`startup.m, disp`

This section provides an introduction to the programming environment in MATLAB. As
mentioned earlier, MATLAB functions are just regular text files containing MATLAB commands. A function text file always has extension .m. For this reason, these files are usually
referred to as m-files. Any standard text editor can be used for developing m-files, and there
is no need to compile m-files. There are mainly two situations where you want to create a
script file:

- To save function calls which take time to find out, when you most likely will need to
  repeat these computations at a later time.

- To save repetitive function calls which take time to type in MATLAB, to save typing
  time in the current or future MATLAB sessions.

A guided tour 15 (Motivation – Why script-files?)

As a first example, suppose you have derived
the following two equations for computing the
braking time and distance for a vehicle with
velocity v having friction μ:

$$v - \int_0^t \mu g \, dt = 0 \Rightarrow t = \frac{v}{\mu g}$$

$$s = \int_0^t v(t) dt = \frac{v}{2} t = \frac{v^2}{2\mu g}.$$

What we can do is to put the MATLAB commands in a text file, exactly as they are
typed in the MATLAB window (without the
prompter).

Typing the name of the script file now works
exactly as copying the text in the file and pasting it into the MATLAB window. All printouts
from the file are shown in the MATLAB window.

```
>> g=9.8;          %[m/s^2]
>> mu=1;
>> v=25;           %[m/s]
>> t=v/mu/g        %[s]
t =
    2.5510
>> d=v^2/2/mu/g %[m]
d =
   31.8878
```

File name: `brake1.m`
```
g=9.8;          %[m/s^2]
mu=1;
v=25;           %[m/s]
t=v/mu/g        %[s]
d=v^2/2/mu/g %[m]
```

```
>> brake1
t =
    2.5510
d =
   31.8878
```

A text editor tailored for MATLAB programming is started inside MATLAB with the
command `edit` (introduced in R11). There are basically three ways to generate the text
file:

1. One can cut and paste the functions into the editor from the MATLAB window. The
   standard short commands work, as **C-c**, **C-v** for copy and paste.

2. One can start a diary by issuing the command `diary brake1.m`. All subsequent commands entered into the workspace will be logged, after typing `diary off` all commands will be saved in a file called `brake1.m`, which you can edit and save for later retrieval.

3. The most common way is to start typing into the editor directly from scratch.

As another example, suppose you are assigned to write a time report every week, where you want to illustrate how the working hours are split between different activities in different ways. We may for instance use the graphical functions `stem`, `bar`, `pie`. Next week, when the new report is going to be prepared, we notice that we should have saved exactly how we did the report last week, and preferably in such a way that it is easy to generate the new report. After all, it is the matrix T that differs between the weeks.

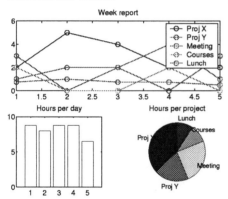

The idea is that when the file name is typed in the MATLAB window, all functions inside the file are executed just as they would have been written in the MATLAB window.

These examples also illustrate two different strategies for inputting data into a script file: In the first example, the velocity and friction are defined in the file. To evaluate the formulas for different values, you have to edit in the file. In the second example, the data vector is defined in the workspace in MATLAB, which is accessible from a script file, and then all plots are generated automatically.

```
>> T=[2 3 1 2 0.75;
      5 0 2 0 1;
      4 2 2 0 0.75;
      2 0 4 2 0.75;
      2 3 1 0 0.5];
>> proj={'Proj X','Proj Y',...
   'Meeting','Courses','Lunch'};
>> subplot(2,1,1)
>> plot(T,'-o')
>> legend(proj)
>> title('Week report')
>> subplot(2,2,3)
>> bar(sum(T'))
>> title('Hours per day')
>> subplot(2,2,4)
>> pie(sum(T),proj)
>> title('Hours per project')
```

File name: `reportfig.m`

```
proj={'Proj X','Proj Y',...
   'Meeting','Courses','Lunch'};
subplot(2,1,1)
plot(T,'-o')
legend(proj)
title('Week report')
subplot(2,2,3)
bar(sum(T'),'w')
title('Hours per day')
subplot(2,2,4)
pie(sum(T),proj)
title('Hours per project')
```

```
>> T=[2 3 1 2 0.75;
      5 0 2 0 1;
      4 2 2 0 0.75;
      2 0 4 2 0.75;
      2 3 1 0 0.5];
>> reportfig
```

Let us return to the braking distance example. Next time you want to compute a braking distance, you will simply search for your script file **brake1.m**. Most likely, you find that it can be improved. For instance, you may want to complement the equations with a nice plot of how the braking distance and time increase with velocity for different friction levels.

To proceed, start by making a copy of the file **brake1.m**, and save it as *e.g.* **brake2.m**. Modify the velocity to a vector on the interesting interval, here ranging from 0 to 50 m/s = 180 km/h $\approx$ 110 mph. Note that the formulas need to be modified for vector calculations by changing * to .* and v^2 to v.^2. To get both functions illustrated in the same figure, **plotyy** is used, which creates two y-axes.

File name: **brake2.m**

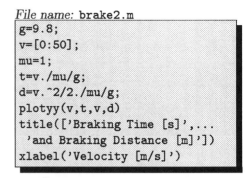

```
g=9.8;
v=[0:50];
mu=1;
t=v./mu/g;
d=v.^2/2./mu/g;
plotyy(v,t,v,d)
title(['Braking Time [s]',...
  'and Braking Distance [m]'])
xlabel('Velocity [m/s]')
```

`>> brake2`

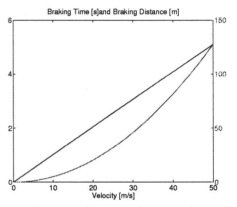

It is crucial that all the functions you develop have the extension `.m`, otherwise they will not be recognized by Matlab. Note here that some text editors save text files with a default extension, like the Windows Notepad using `.txt`. There are some other issues related to how Matlab searches for m-files that need to be solved before you can start extending your Matlab system with your own functions.

A guided tour 16 (The Matlab path)

The Matlab commands **pwd** and **dir** were illustrated in Section 4. These commands display the path to the current working directory and lists the files in this directory, respectively. The command **cd** changes the current working directory, either relative to the current one or to a specified directory. Any command following an *exclamation mark* ! is issued to the underlying operating system of the computer.

In the example above, a new directory is created using the operating system command **mkdir**. There is actually a Matlab command with the same name having the same effect.

```
>> pwd
ans =
C:\matlabr12\work
>> !mkdir project
>> cd project
>> dir
.                    ..

>> cd ..
>> mkdir lib
>> dir
.                    project
..                   lib
```

When searching for a function **fun**, MATLAB first looks in the current working directory. Thereafter, it searches for the first match of the sought file **fun.m** in the MATLAB search path. Issue the command **path** to view your current MATLAB search path. You can extend the search path with your own directories. The command **addpath** prepends a specified directory to the MATLAB path. There is also a corresponding command **rmpath**.

```
>> addpath C:\matlabr12\work\lib

>> rmpath C:\matlabr12\work\lib
```

However, using **addpath**, the extended search path is only valid for the current MATLAB session. The path definitions have to be saved on the computer disk and rerun for each MATLAB launch. There is a convenient **pathtool** for editing the MATLAB path (introduced in R11). When choosing **Save Path** from the **File** menu in this interactive program the file **pathdef.m** is updated with the changes made. This file is automatically executed when MATLAB is started.

```
>> pathtool
```

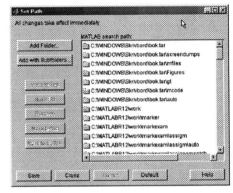

Since MATLAB searches for the first occurrence of the sought m-file it is important to have the path definitions in the right order. If there is a risk for name conflict, it is convenient to use the command **which** to list the search path to the command.

```
>> which pathdef
C:\matlabr12\toolbox\...
    local\pathdef.m
```

Instead of using the **pathtool** command to edit the search path one can personalize the path and several other aspects of MATLAB using a personal startup file. If there is an m-file with name **startup.m** on the initial search path during MATLAB launch, it will automatically be executed. You can test if such a file is present on your MATLAB installation by issuing the command **which startup**, you will either receive an error message or the complete search path to this file. If there is no such file defined on your system, you can create a new file and use it to personalize your installation of MATLAB. On multi-user platforms like Unix, the startup file is most commonly placed in your home directory, or in a subdirectory called **matlab**. On single user systems it can be placed anywhere on the search path, it is however advisory to save the file in the working directory used for MATLAB startup. This working directory is displayed by issuing the command **pwd** directly after MATLAB has been launched.

A guided tour 17 (Personal startup file)
Create an empty text file called `startup.m` and save it somewhere on the MATLAB search path. This can be achieved by making sure that the current work directory is on the search path and starting the built-in m-file text editor issuing the command `edit`.

```
>> edit
```

The startup m-file could for example include changing the output format, changing the directory to an appropriate one, displaying a brief message, and extending the matlab path to a home made library of written m-files. The command `disp` displays the array argument without printing the array name, in this case it displays a concatenated character string.

File name: `startup.m`

```
% My personal startup file
format compact
cd c:\matlabr12\work\project
disp(['Working directory ',pwd])
addpath c:\matlabr12\work\lib
```

Save the startup file under the name `startup.m`, quit matlab and restart it again to verify that the startup file is executed automatically. Some useful settings for `startup` will be given in guided tour 34.

When you quit MATLAB with `quit` or `exit`, the file `finish.m` is executed. This file can also be modified, for instance by always saving workspace to avoid losing data by mistake.

On many platforms, the editor is also available from the **File** menu in the MATLAB window. Choose **New** and then **M-File** or click on the empty white paper icon.

When using MATLAB to solve engineering problems it is convenient to extend the MATLAB functionality with personal m-files, save data in mat-files, and perhaps save illustrations as fig-files. In order to run these m-files and retrieve the saved data you either have to change the current working directory to the location of these files, or you have to make sure that the MATLAB search path points to their location. The latter procedure is usually more appropriate and can be implemented using a startup file as in the example above.

Exercise 42
Write a script file `addpwd` that adds the working directory to the search path.

Exercise 43
Down-load the m-files `tetris.m` and `fern.m` from the book home page

   `http://www.control.isy.liu.se/books/matlab`

and place them in a suitable newly created directory. Update the file `startup.m` so that you have access to these new files, restart MATLAB and run the command `tetris`. Ignore the warnings if you are running MATLAB version 5 or higher. Enjoy!

9   MATLAB **Functions**

Content: The MATLAB programming language.	
Flow control:	if, else, end, for, while, switch, case, break
Functions:	break, input, %, keyboard, dbstop, return
	error, nargin, nargout
Variable context:	global, persistent
Timing:	clock, etime, cputime, tic, toc

There are two types of program files in MATLAB: *script-files* and *functions.* These are both referred to as m-files, since they need to have the file extension .m. The only basic difference between them is how they treat variables. A script file uses the global variables defined in the MATLAB workspace. Hence, running a script file is equivalent to executing a sequence of command lines in the workspace. In contrast, a function m-file has local variables and the input and output parameters need to be specified. The startup file **startup.m** given in Section 8 is an example of a script file.

The differences between script m-files and function m-files are summarized in the table below.

script m-files	function m-files
No input or output arguments.	May take input and deliver output arguments.
Operates on the variables in the workspace. Variables declared in the script will be available in the workspace after execution of the script.	Operates on local variables by default. May operate on workspace data if variables are declared as **global** and on previous local variables if declared as **persistent**. May have sub-functions.
Main use for automation of a series of command steps that needs to be re-entered many times during interactive development in the workspace, for example a common combination of plotting and visualization commands.	Mainly used to extend the MATLAB system with new commands from your own application field.

Table 9.2 The difference between a script file and an m-file.

An m-file looks like a program in any common language, like C, Fortran, Pascal, Basic or Java. However, there is no need for variable declaration or compilation. The great advantage is that it is quite simple to write your own programs in MATLAB, view, edit and print them using any text editor. Moreover, the m-files are *platform independent* in general, and are executable on any platform running MATLAB.

A guided tour 18 (script vs. function)

Let us return to the first script example in guided tour 15. The most obvious advantage with making a function of this script is that the input variables are entered in a natural way, and the output is returned just as in any other function like y=sin(x). Make a copy of the file brake1.m and save it as *e.g.* brake3.m. Note that the variables t,d are local in a function, that completely different names can be used in the function call, as braketime here, and that not all output variables need to be returned.

Consider a function that computes the factorial $n! = 1 \cdot 2 \cdots n$. An example script m-file named fact.m is shown to the right. A MATLAB script is completely equivalent to running the sequence of commands directly from the command prompt. There is no need to compile the m-file or declare any variables before usage. Once the m-file is stored somewhere in the MATLAB path it can be called from MATLAB. Note that any predefined variable y in the workspace will be redefined after running the fact command. This problem is circumvented by redefining fact.m to a function m-file with local variables instead.

All function m-files start with a function header indicated by the identifier function followed by the function syntax definition. The syntax definition serves to name the function and the input and output parameters. The file name, in this case fact.m, and the name in the function header should always be the same. Note that when calling the function there is no need to define a specific input variable n in the workspace. The semicolon ; is used to suppress the display of the assignment to y inside the function. Note also that the variable y inside the function is a variable with local scope. The variable y defined in the MATLAB workspace is a completely different variable. The output from the function is copied to the variable nfac used during this particular call.

File name: brake3.m

```
function [t,d]=brake3(v,mu)
% Computes braking distance d
% and braking time v from speed
% v and friction mu
g=9.8;          %[m/s^2]
t=v/mu/g;       %[s]
d=v^2/2/mu/g;   %[m]
```

```
>> [braketime]=brake3(25,1)
braketime =
    2.5510
```

File name: fact.m

```
% Factorial y = 1*2*..*n = n!
y = prod(1:n);
```

```
>> n = 10;
>> fact
>> y
y =
    3628800
>> n = 5; fact, y
y =
    120
```

File name: fact.m

```
function y = fact(n)
%Y=FACT(N) Factorial n!=1*2*..*n
y = prod(1:n);
```

```
>> nfac = fact(6)
nfac =
    720
>> y
y =
    120
>> y = fact(6)
y =
    720
```

The first consecutive lines with comments are displayed when issuing the **help** command on the function. Note that it is customary to include the calling syntax in one of the first rows and that the first row should contain a one-line description of the function containing the most relevant keywords. The function **lookfor** looks for a keyword in the first help line in all functions in MATLAB's search path.

```
>> help fact
 Y=FACT(N) Factorial n!=1*2*..n
>> help brake3
 Computes braking distance d
 and braking time v from speed
 v and friction mu
>> lookfor braking
brake3.m: % Computes braking...
           distance d
```

Let us take a closer look at the function and script m-file examples given above. There is a general pattern that is common to all MATLAB m-files, and it divides the m-file into four distinct parts. This is detailed in the commented example of a more elaborate implementation of the function **fact**, shown below.

File name: **fact.m**

```
function y = fact(n)           % function definition line
%Y=FACT(N) Factorial n!=1*2*..*n   % H1-line
%FACT(N) computes the factorial    % Help text
% of N, usually denoted N! Only
% valid for integer, positive N

% The function body (this comment is not part of the help text)
y = prod(1:n);
```

Function definition line The first line in the m-file defines the function name and the order and number of input and output arguments.

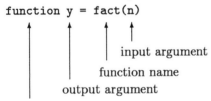

```
function y = fact(n)
```

keyword · output argument · function name · input argument

All MATLAB functions have a function definition line that follows the pattern above. If the m-file does not start with the identifier **function**, MATLAB will treat the m-file as a script.

Functions may have none, one, or any number of output and input arguments. It may even have an undefined number of input and output arguments, where the number may differ each time it is called. Multiple input arguments are enclosed in parentheses and output arguments are enclosed in brackets, commas are used to separate each argument

```
function [x,y,z] = cylinder(r,n)
```

A function may be *called* with any number of arguments as long as it does not exceed the number in the function definition line. The variables that are passed to the function and returned from the function naturally need not have the same name as given in the function definition line. A valid call to **cylinder** can look like

```
>> [zz,ff] = cylinder(10);
```

the function body can be written in such a way that the function behaves differently depending upon the number of input and output arguments that were used when calling the function.

H1 Help text line The first non-empty line following the function definition line is written as a comment briefly describing the function. Everything on a line in an m-file following a percent sign % is treated as a comment by MATLAB. This line is searched and displayed when using the command `lookfor` to find a specific MATLAB function. This line is also displayed when issuing the `help` command on the directory that the m-file is located. Since this line is searched by the `lookfor` command it is important to keep it as descriptive as possible. Note that this line is also displayed in the list generated by `help dir`. Try `help(pwd)` for instance.

Help text The preceding comments are displayed together with the H1-line, when calling for help on this specific m-file. This help information can contain any information about the m-file. Usually it describes the functionality provided by the m-file and the proper usage of the function. The different input and output arguments are also described in detail here. The help text should be informative and exhaustive so that there is no need to read the m-file code to understand the function. However, the help text should not be too elaborate since this information should preferably fit on one screen of the command window. (For some functions with many options, like `plot`, one has to turn off *scrolling* by `more on` to see all the help text.)

M-file body The first line following the function definition line which is not started by a *comment character* % will determine the start of the function body. In the example above the function body is distinguished from the help text by an empty line. The function body contains the actual code that performs the calculations and assigns values to any output variables. The statements in the function body may consist of calls to other functions and scripts, different programming constructs like flow control and interactive input and output, calculations, assignments, comments and blank lines. It is advisable to end all lines in the function body with a semi colon ; to suppress any printout of results of computations performed in the m-file.

MATLAB m-file names are constrained similarly to the way variables may be named in MATLAB. Note though, function names are not case sensitive in contrast to variables. It is common to refer to m-files using lower case characters.

Technically, the first 31 characters of the function name is used to distinguish the m-file. The name must begin with a letter, and may be followed by any letter, number or underscore character. The text file that contains the m-file is named as the m-file name with the extension `.m` appended. In the example above, the text file is called `fact.m`. The file name will determine the name of the function if the name on the function definition line is different from the m-file name. It is, however, strongly recommended that these two names are the same.

Logical Relations

We start with defining the most useful logical relations, since these are fundamental for controlling program flow.

A guided tour 19 (Logical relations)

To understand the logical program constructions to follow, we start by examining relational operations. By a relation operator is meant an expression that can be either 'false' or 'true'. MATLAB represents true by 1 and false by 0. Comparisons like >,<,==, = are the most common examples. Note that a double equal sign is used to distinguish assignments and relations (try a=1, a==2, a=2, a==2).

The summary on page 12 is repeated here for convenience.

<	less than	&	logical and	
>	greater than	|	logical or	
<=	less than or equal	~	logical not	
>=	greater than or equal	xor	exclusive or	
==	equal	1	true	
~=	not equal	0	false	

There are also some useful functions that operate on vectors and matrices, like **any, all**. Here all logical relations interpreted element-wise in the vector or matrix must be true.

Finally, note in the example how to test if something is equal to the empty set.

```
>> 3>2
ans =
     1
>> 3<2
ans =
     0
>> 3==2
ans =
     0
>> 3~=2
ans =
     1
>> 3~=2 & 3>2
ans =
     1
>> any(3>[1 2 3 4])
ans =
     1
>> all(3>[1 2 3 4])
ans =
     0
>> a=[];
>> a==[]
Warning: X == [] is technically
incorrect. Use isempty(X) instead.
ans =
     1
>> isempty(a)
ans =
     1
```

Controlling Program Flow

Several fundamental *programming structures* exist to control the program execution flow. These flow control statements generally consist of selection and repetition, or any combination of these two. The first fundamental flow control construct is the *selection statement* if. The syntax is given by

```
if <logical condition>
   <statements>
elseif <logical condition>
   <second case>
else
   <otherwise>
end
```

If the logical condition is true, MATLAB executes all statements between the **if** and **end** lines. Execution is resumed at the line following the **end** statement. If the logical condition is false, the statements between **if** and **end** are skipped and the execution resumes at the line following the **end** statement.

A guided tour 20 (Selection using if)

As a first example, we will write a function that removes files from the disk using **delete**, but first checks if there is such a removable file or not. For this purpose, the function **exist** can be used. It returns 0 if its argument is not a variable or file in MATLAB's search path, and 2 if it is a file. The other cases are of no interest in this example.

The **if** statement in the function **remove** first checks if the argument is a file. In such case, the file is removed and a message is displayed. If it is not a file, the **elseif** condition is tested. If it is true, which happens when no such file or variable exists, another message is displayed. If neither of the **if** or **elseif** conditions are true, the last message is shown.

File name: `remove.m`

```
function remove(file)
b=exist(file);
if b==2
   delete(file)
   disp([file,' is removed!'])
elseif b==0
   disp([file,' does not exist'])
else
   disp([file,' exists but ',...
   'is not removable'])
end
```

```
>> remove('sin')
sin exists but is not removable
>> remove('temp.m')
temp.m does not exist
>> remove('tmp.m')
tmp.m is removed!
```

Next, we return to the factorial example, and try to make the function more user-friendly. Introduce a test to verify that the function has been called with a valid input. The command **nargin** determines the number of input arguments used when calling a MATLAB function. Similarly, **nargout** replies the number of output arguments used in the call. If the input **n** is invalid, the program reaches an **error** statement which will display the error message and halt the program. The example to the right shows how several conditions can be nested using the **elseif** statement.

File name: `fact.m`

```
function y = fact(n)
%Y=FACT(N) Factorial n!=1*2*..n
if nargin < 1
   error('no input assigned')
elseif n < 0
   error('input not positive')
elseif abs(n-round(n))>eps
   error('input must be integer')
end
y = prod(1:n);
```

```
>> fact(3.4)
??? Error using ==> fact
input must be integer
>> fact
??? Error using ==> fact
no input assigned
```

There is an alternative selection control statement called **switch** (introduced in R11), illustrated in the script m-file to the right. The **input** function prompts the user for input and stores the result in a character string. The general syntax for the switch statement is

```
switch <switch_expr>
  case <case_expr>
    <statements>
  otherwise <case_expr>
    <statements>
end
```

The `otherwise` statement is optional.

A guided tour 21 (Selection using `switch`**)**
The function `exist` might be improved to be
more explicit, since there is no way for the
user to remember all cases and getting `help`
is necessary. The modified file to the right
displays the type of file or variable for the most
common file types.

File name: `myexist.m`

```
function myexist(name)
switch exist(name)
case 0
  disp('not found')
case 1
  disp('variable')
case 2
  disp('file')
case 5
  disp('built-in function')
case 7
  disp('directory')
end
```

Multiple cases can be grouped together by en-
closing them in curly brackets. The statement to the right illustrates this, assuming
that there is a scalar variable called `choice`
available.

```
switch choice
  case {1,4}
    disp('1 or 4')
  case {2,3}
    disp('2 or 3')
end
```

This example shows a sophisticated way to
quit MATLAB. It implements an interactive
control question, with a recursive call to the
function itself after a random waiting time until 'Yes' is answered. The function `input` is
useful for interactive questions, and `upper`, or
`lower` for conversion to uppercase or lower-
case letters, implying that the user can answer
'YES', 'yEs' or any other combination.

File name: `windas.m`

```
%Windas too thousand for Aussies
v = input('Are you sure? ','s');
switch upper(v)   % Upper case
case 'YES'
  quit
case 'NO'
  pause(5*rand(1))
  windas
otherwise
  pause(5*rand(1))
  disp('Answer yes or no! ')
  windas
end
```

Note that, contrary to C, the switch case does not fall through: only the statements between the matching **case** and the following one are executed. Instead of comparing strings, the switch expressions can also be any logical expression and the case any scalar value.

```
>> windas
Are you sure? No
Are you sure? No, I said!
Answer yes or no!
Are you sure? Yes
quit
```

The condition statements determine which part of the code to execute depending on some logical expression. Another cornerstone in program flow control is repetition which enables you to write compact code for executing repeated sections of identical, or almost identical, code. The general repetition syntax is given below.

```
for k=<vector>
    <statements>
end
```

A guided tour 22 (Repetition using for)

As a first example, we want to simulate a so called chaotic system defined by the recursion

$$x(i) = \lambda x(i-1)(1 - x(i-1)),$$

for $\lambda \in [3.8, 4]$ and $0 < x(1) < 1$. An example of a few iterations is shown to the right. Clearly, this manual repetition is not suitable for computing for instance $x(1000)$.

```
>> x=0.9;
>> x=3.9*x*(1-x)
x =
    0.3510
>> x=3.9*x*(1-x)
x =
    0.8884
>> x=3.9*x*(1-x)
x =
    0.3866
```

This is a typical example of a repetitive calculation suitable for a **for** loop. This program automatically repeats the statement $x(i) = lx(i-1)(1 - x(i-1))$. This is done n times. Another advantage here is that the sequence $x(i)$ becomes indexed, enabling the possibility to plot the sequence.

File name: `chaos.m`

```
function x=chaos(x0,l,n);
% Simulation of x=l*x*(1-x)
x(1)=x0;
for i=2:n
    x(i)=l*x(i-1)*(1-x(i-1));
end
```

```
>> x=chaos(0.9,3.9,4)
x =
    Columns 1 through 4
    0.9000   0.3510   0.8884   0.3866
```

Any program flow command can be written in the MATLAB window as a little program in itself. Sometimes it is handy to execute a repetition directly in MATLAB without creating a file. This example prints out a table of cumulative sums $\sum_{i=1}^{n} i$ and products $\prod_{i=1}^{n} i = n!$. Note that when writing program inside MATLAB, the prompter is suppressed until the last **end** appears. The repetition starts after entering the return key after the last **end**.

A repetition statement is illustrated to the right by an example of testing if a number is prime. A straightforward way to determine if a given integer number n is prime is to test if any of the integers less than its square root is a divisor to n. The function **rem** determines the remainder after integer division.

```
>> for n=1:6;
    disp([n sum(1:n) prod(1:n)]),
  end
       1       1       1
       2       3       2
       3       6       6
       4      10      24
       5      15     120
       6      21     720
```

File name: `primecheck.m`

```
function p = primecheck(n)
%P = PRIMECHECK(N)
%    P == 1 if N is prime.
p = 1; % Assume prime
for k=2:sqrt(n)
   if rem(n,k)==0
       p = 0;
   end
end
```

```
>> primecheck(9973)
ans =
       1
>> for e = eye(3)
  e'
end
ans =
       1       0       0
ans =
       0       1       0
ans =
       0       0       1
```

The fixed length repetition statement **for** assigns each entry sequentially to the repetition variable (in the case above **k**) and executes the statements inside the repetition scope. The index list for the repetition variable may be any matrix with **m** rows and **n** columns. The repetition variable will then be assigned each column of this matrix sequentially, and the loop will thus be executed **n** times. In the illustrative example to the right, the variable **e** is a vector containing one of the columns of the unit matrix **eye(3)** during each repetition of the loop.

Repetition loops in MATLAB should in general be avoided in favor of the vectorized notation. Applying elementary functions entry-wise on matrices is a lot faster than looping through the matrix entries. The example without loops given to the right indicates that this can be both a compact and efficient way of programming. Refer to the help text of **rem** for details regarding its vectorization.

File name: **primechk.m**

```
function p = primechk(n)
%P = PRIMECHK(N)
%    P == 1 if N is prime.
p = ~any(rem(n,2:sqrt(n))==0);
```

Execution speed can be measured using the MATLAB stopwatch timer. The command sequence **tic, command, toc** prints the time for executing **command**, see also **clock**, **etime** and **cputime**. One should in general avoid using long repetition loops in MATLAB and instead try to use a vectorized programming scheme like the one exemplified in **primechk.m** above. For this example, there is a built-in function **isprime** in MATLAB that computes exactly the same thing. Generally, compiled C-code is even faster than vectorized solutions. See also Chapter 14.

```
>> tic, primecheck(499999997); toc
elapsed_time =
    0.1700
>> tic, primechk(499999997); toc
elapsed_time =
    0.0200
>> tic, isprime(499999997); toc
elapsed_time =
    0.0100
```

An alternative to the **for** repetition is obtained by the *repetition statement* **while**. The syntax is given by

```
while <logical condition>
  <statements>
end
```

It is quite useful for numerical computations, where one is not sure exactly how many repetitions are needed.

A guided tour 23 (Repetition using while)
To find the largest factorial MATLAB can compute, a first try is to use a very long **for** loop, and using **return** to terminate once Inf is returned. We re-use the function **fact.m** and write the function **factlimit** to the right.

File name: **factlimit1.m**

```
% factlimit finds the largest
% integer n for which n!<realmax
for i=1:10000
  if fact(i)==Inf
    n=i-1
    return
  end
end
```

```
>> factlimit1
n =
   170
```

Note that this first implementation required both a repetition **for** and a condition **if**. Program flow control using **while** combines the repetition and condition. Though **factlimit1** and **factlimit2** give the same answer, the second implementation is clearly to prefer.

File name: `factlimit2.m`

```
% factlimit finds the largest
% integer n for which n!<realmax
n=1;
while fact(n+1)<Inf
    n=n+1;
end
n
```

```
>> factlimit2
n =
    170
```

We next give a more realistic example of a program solving a problem of more general interest. *Newton's method* is a famous numerical algorithm for finding roots to nonlinear equations, $f(x) = 0$. Starting with an initial guess x_0 the method follows the tangent to the nonlinear curve at $f(x_0)$ which yields the recursive expression

$$x_{n+1} = x_n - \frac{f(x_n)}{f'(x_n)}$$

This method is in general quite fast, but there is no guarantee that it converges.

The function **newton** to the right implements the recursion according to Newton's method for the particular case $f(x) = e^{-x} - \sin(x)$, and repeats it until no further significant improvement is achieved or until the maximal number of iterations is exceeded. Note that a command line can be extended over several physical lines using three subsequent dots. Note also the syntax for defining functions with several output variables using brackets.

File name: `newton.m`

```
function [x,n]=newton(x0,ep,mn)
%[X,N]=NEWTON(X0,TOL,MAXIT)
x = x0;
n = 0;
while (abs(x-x0)>ep|n==0)&n<mn
    n = n + 1;
    x0 = x;
    x = x - (exp(-x)-sin(x))/...
        (-exp(-x)-cos(x));
end
```

```
>> [x,n]=newton(0,1e-3,10)
x =
    0.5885
n =
    4
```

The Newton method illustrates a flow control statement using both repetition and condition, having the general syntax:

```
while <logical condition>
  <statements>
end
```

The program repeatedly executes the statements inside the repetition scope as long as the logical condition is true.

Starting at $x_0 = 0$, convergence within an absolute tolerance of 10^{-3} is achieved in four iterations, as illustrated to the right.

```
>> t = 0:0.01:1;
>> plot(t,exp(-t)-sin(t),'-',...
       x,0,'o')
>> legend('f(x)',...
       'numerical root')
>> grid
```

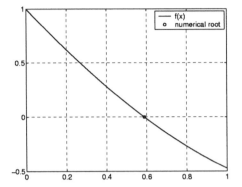

The flow control structures demonstrated above can be nested in any order relative to each other. This yields a great programming flexibility and problems can often be solved in several ways using these fundamental control structures. Still, there may be cases that are hard to solve by nesting repetition and selection statements. Then, the **break** command in combination with **if** is often the solution. If a **break** statement is executed inside a repetition loop, the loop is terminated prematurely. In nested loops, only the innermost loop is terminated. If a **return** statement is executed anywhere in an m-file, the m-file is terminated.

To illustrate efficient programming, the task here is to compute a list of prime numbers. The prime numbers are stored in a vector p, and a loop of all integers up to a given upper limit is performed. For each value of the loop variable, the program checks if the loop variable is divisible with any of the previously found prime numbers. If not so, the list p is extended.

In this first implementation, a second loop over all elements in the vector p is performed. This is perhaps the most straightforward solution.

File name: **prime1.m**

```
function p=prime1(n);
% List all prime numbers up to n
p=2;
for i=3:n
  primeflag=1;
  for ptest=p
    if i/ptest==round(i/ptest)
      primeflag=0;
    end
  end
  if primeflag
    p=[p i];
  end
end
```

Utilizing *vectorization* in MATLAB programming is traditionally the single most important factor for performance. However, MATLAB R13 introduced so called *performance acceleration* and the explicit goal is to make MATLAB as fast as C or Fortran. At this stage, R13 offers increased speed in self-contained **for** loops (not calling any m-files inside the loop).

In the second more optimized program, the integer i is divided by all elements of p at the same time, and the remainder is computed by the function rem. If any remainder is zero, the integer i is divisible with at least one prime number in p. In other words, if there is not any remainder that is zero, we have found a new prime number. Remember that the tilde ~ is used for negation in MATLAB.

File name: `prime2.m`

```
function p=prime2(n);
% List all prime numbers up to n
p=2;
for i=3:n
  if ~any(rem(i,p)==0)
    p=[p i];
  end
end
```

To understand the vectorized solution, try

```
p=[2 3 5 7 9 11];
rem(13,p),            rem(15,p)
rem(13,p)==0,         rem(15,p)==0
any(rem(13,p)==0),  any(rem(15,p)==0)
```

To compare the implementations, we execute them both and use tic,toc for timing them. The difference increases with the input argument. The main difference of the programs is that the latter one implements the inner loop in built-in C-code, which is much faster than a MATLAB loop.

Next, we use the second implementation to compute a list of all prime numbers up to 10000 (it takes just a few seconds), and a *histogram* over the difference between two consecutive prime numbers is presented. It rarely happens that it takes more than 30 integers before the next prime number appears.

To further increase performance, see the MATLAB function **primes**. It uses a different principle. Start with a list with all odd numbers as potential prime numbers, and then delete the ones that are multiples of each new found prime number. Enter **type primes** and try to understand it, see Section 14 in Part 2 for a brief explanation of the code.

```
>> prime1(20)
>> tic,prime2(100);toc
elapsed_time =
    0.0367
>> tic,prime1(100);toc
elapsed_time =
    0.0530
>> p=prime2(10000);
>> hist(diff(p),1:40)
>> xlabel('Prime distance')
```

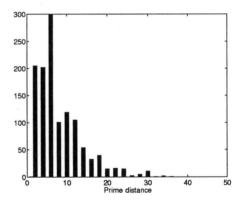

Sub-functions

A MATLAB function may use *local functions* which are accessible only within the function and thus not from the command window. This is similar to the use of local variables in functions. The first function in the m-file is called the *primary function*. Additional functions are called *sub-functions*, these are only visible to the primary function and other sub-functions in the same m-file.

A guided tour 24 (Sub-functions)

Sub-functions are convenient for separating computations in the m-file. In the example to the right, there are two sub-functions used to compute the median of a vector of data. The median of a set of numbers is defined as the number splitting the set in half, having as many larger as smaller numbers in the set. The internal functions **odd** and **even** are not accessible from outside the m-file **median.m**.

File name: `median.m`

```
function mu = median(u)
% MEDIAN(U) Compute the median
us = sort(u);
n = length(us);
if rem(n,2) == 0
  mu = even(us,n);
else
  mu = odd(us,n);
end

function y = odd(x,n)
y = x((n+1)/2);

function y = even(x,n)
y = (x(n/2) + x(n/2+1))/2;
```

Name resolution, elaboration and arguments

In MATLAB we use parenthesis () both for indexing matrices and when calling functions with input parameters. This can sometimes cause confusion since MATLAB may interpret a function call as a matrix indexing. When MATLAB runs across an identifier in an expression, it searches for the correct interpretation of this identifier and runs through the following steps:

1. Check if the name is a variable.

2. Check if the name is a built in function.

3. Check if the name is a sub-function.

4. Check if the name is a *private function*. A private function is stored in a directory named **private** and it is only accessible to functions in the directory immediately above this directory.

5. Check if the name is a function that resides somewhere on the MATLAB search path. The first matching function in the search path will be evaluated.

If there are duplicate names the first applicable rule in the ordered list above will be chosen.

Debugging

A convenient tool for *debugging* m-files is the command **keyboard**. Executing a **keyboard** statement in an m-file, the execution is stopped and the prompt K>> appears in the command window. The local variables defined inside the function can then be investigated for inconsistencies, and changed, from the command prompt K>>. Issuing the command **return** will cause MATLAB to continue the execution of the m-file at the position it was aborted.

Another debugging possibility is to issue the command

```
>> dbstop if error
```

prior to running the program. Instead of aborting the program execution when an error occurs, it invokes the keyboard just as if **keyboard** had been written on the line the error occurred. Once again, **return** will cause the program to continue its execution where aborted. See **help dbstop** for several additional options. On later versions of MATLAB, the m-file editor **edit** will automatically be started and an arrow indicate the position where the error occurred. The debugger state is reset by issuing the command **dbclear** from the command line. The debugger state and manual break points can also be edited directly inside the m-file editor under the **Debug** menu.

Exercise 44
Inspect the hilb *function and try to understand how the colon operator is used to avoid the, at first glance, unavoidable loops.*

Exercise 45 (The hailstorm series)
Let $n > 1$ be an integer and form a sequence where the next number is $3n + 1$ if n is odd, and $n/2$ if n is even. A famous unproven conjecture says that whatever positive integer n we start with, the sequence will always reach 1. Write a program that computes the sequence from n to 1. The program should test that the input argument is an integer $n > 1$ if not, the program should halt and report an appropriate error message. Test example:

```
>> a = hailstorm(6)
a =
     6    3   10    5   16    8    4    2    1
```

Use your program to plot the number of iterations needed to reach 1 versus the starting number for $1 < n < 10$.

Exercise 46
Write a program with one input n that finds the best rational approximation $\frac{p}{q} \approx \pi$ where both integers p and q have at most n decimal figures. Syntax illustration and test example:

```
>> [p,q] = ratpi(3)
p =
   355
q =
   113
```

Exercise 47
Write a rational approximation program having two inputs that can find a rational approximation to any given real number x up to a given relative error

$$\frac{|x - \frac{p}{q}|}{|x|} < \varepsilon$$

Use your program to find good rational approximations to π, $\sqrt{2}$, and e^1. Syntax and test example:

```
>> [p,q] = ratapprox(pi,1e-3)
p =
    22
q =
     7
```

Exercise 48 (Fibonacci numbers)

Design a function that computes the n'th Fibonacci number f_n, defined by

$$f_0 = 0, \qquad f_1 = 1, \qquad \text{and} \qquad f_n = f_{n-1} + f_{n-2} \qquad \text{for } n > 1.$$

Syntax: y=fib(n). There are three obvious implementations of this algorithm

Iteration Straightforward implementation based on a loop from 3 to n.

Recursion Starting backwards at n calling itself using the the third equation above. Return the values 0 and 1 when $n = 0$ and $n = 1$, respectively.

Vectorized View the recursion as a linear time invariant discrete dynamical system

$$\begin{pmatrix} f_n \\ f_{n-1} \end{pmatrix} = F \begin{pmatrix} f_{n-1} \\ f_{n-2} \end{pmatrix}$$

for certain choices of the constant matrix F and initial state defined by f_0 and f_1. How can you efficiently compute the output of this system for any $n > 1$?

Compare the computational burden by timing your functions using very large values of n.

Exercise 49 (Square root)

Use the following more than 2000 year old method to implement a square root algorithm. Let y denote the square root of x, that is $y^2 = x$. If y_0 is a guess of the square root, then the mean of y_0 and x/y_0,

$$\frac{y_0 + x/y_0}{2},$$

is a better guess. The syntax should be y=mysqrt(x,yinit,tol) where the iterations terminate when two consecutive guesses are closer than tol.

Exercise 50 (McNugget numbers)

The McNugget numbers are defined by the possible number of chicken McNugget's you can order at McDonald's. Remember that they are served in quantities of 6, 9 or 20. That is, $35=6+9+20$ is a McNugget number. Write a script m-file that computes the largest non-McNugget number. Hint: when you have found six consecutive McNugget numbers, you know that there are no larger non-McNugget numbers (why?).

Exercise 51 (The Mandelbrot fractal)

For every point c in the complex plane we can define the recursion

$$z_0 = 0 + 0i$$
$$z_n = z_{n-1}^2 + c$$

The Mandelbrot set is defined as the points for which $|z_n|$ stays bounded. It can be shown that if ever $|z_n| > 2$, c is not in the set. To visualize the set, we let N be the number of iterations needed to get $|z_N| > 2$. That is, for each c in the complex plane there is an $N = N(c)$ that can be illustrated in different ways. For example, we can let each N correspond to a color, and plot the colors of the points in the complex plane.

  Hints: Start by writing a function n=mandel(realgrid,imaggrid,Nmax), which calculates the number $N(c)$ for each point c in the grid. Remember that iterations are slow in MATLAB and try to write a function that uses the entry-wise matrix computations. You might have need for the command any. There are several ways to visualize the result. The commands contour and pcolor might be useful. The interesting area lies in the square $-2 < Re(c) < 1$ and $-1.5 < Im(c) < 1.5$

10 Functions of functions

> **Content:** Writing general purpose functions operating on other functions.
> **Functions:**
> `quad, quadl, fminbnd, fzero, ode45, eval, feval, inline, @`

Numerical integration can be performed in MATLAB by numerically evaluating the integrand and approximating the integral with a finite sum, see Exercise 18 and Exercise 19 on page 15. This integral approximation scheme lends itself for m-file implementation, having the integration interval defined by some input parameters. However, if we are interested in computing an integral with respect to another integrand, a new m-file needs to be written. For sake of generality, it would be convenient to be able to write general purpose integration functions accepting the integrand as an input parameter. In MATLAB such *functions of functions* can be constructed by delivering the function name in a string as input to the general purpose function.

A guided tour 25 (Standard functions of functions)

General purpose numerical integration is performed by the functions **quad** or **quadl** in MATLAB. The first argument to **quad** is a string containing the name of the function that evaluates the integrand.

```
>> quad('sin',0,pi)
ans =
    2.0000
```

The numerical integral

$$\int_0^1 xe^{x^2}\,dx \approx 0.8591$$

File name: `myfun.m`

```
function y = myfun(x)
y = x.*exp(x.^2);
```

is computed by calling **quad** with a string informing that the integrand is called **myfun**, according to the syntactical expression shown to the right. The file **myfun.m** must of course be written and saved on the MATLAB search path. Note that the code in **myfun.m** must be written in vectorized form, using the operators `.*` and `.^`, respectively.

```
>> quad('myfun',0,1)
ans =
    0.8591
```

Examples of other general purpose functions are **fzero** for finding roots to scalar nonlinear functions, and **fminbnd** for minimization of scalar functions over a bounded interval. In earlier versions of MATLAB **fminbnd** is called **fmin**. Ordinary differential equations are solved with **ode45**. See **help funfun** for more examples.

```
>> fzero('myfun',pi)
ans =
    -2.8500e-18
```

In order to construct general purpose commands we must be able to access a function given its name stored in a character array. The MATLAB function **feval** evaluates a function specified by its name given in a character array, while the function **eval** evaluates a complete MATLAB expression stored in a character array.

A guided tour 26 (Customized functions of functions)

The general purpose function `fdiff` implements a numerical derivation using a fractional difference approximation,

$$f'(x) \approx \frac{f(x+h) - f(x)}{h}$$

```
File name: fdiff.m
function fprime=fdiff(f,x,h)
%fprime=fdiff(f,x)
if nargin < 3
   h = 0.01;
end
fprime = (feval(f,x+h)-...
   feval(f,x))/h;
```

The function `feval` can evaluate functions taking any number of input and output arguments.

Note the way that **nargin** is used to give the third parameter a default value when not specified. A smaller value of this parameter gives a more accurate approximation, closer to the analytical value.

```
>> fdiff('myfun',1)
ans =
    8.2925
>> fdiff('myfun',1,1e-3)
ans =
    8.1685
>> exp(1)+2*exp(1)
ans =
    8.1548
```

Generally, for functions taking another function `f` as input, it can be defined in any of four different ways:

1. The function name given in a string: `f='sin'`,

2. An expression given as a string: `f='(x+1)/(x-1)'`,

3. An *inline object*: `f=inline('(x+1)/(x-1)')`,

4. A *function handle*: `f=@sin`,

Alternative 4 should be the fastest one in time-critical applications, followed by 1. Alternatives 2 and 3 are equivalent, since an inline object is created internally in 2.

A guided tour 27 (Function handles)

The function **humps** is pre-defined in MATLAB. We first create a function handle using the *at* @: character. This gives the same consequences as `f='humps'`, but it is a faster and more flexible alternative.

The minimum of a function can be computed by `fminsearch`. It uses a gradient algorithm, so it is important to initialize the algorithm correctly (here in 0.5). `feval` is used to compute the function value at the minimum.

```
f = @humps;
fplot(f,[-1,2])
xm = fminsearch(f,0.5)
xm =
    0.6370
fm = feval(f,xm)
fm =
    11.2528
hold on
plot(xm,fm,'o')
```

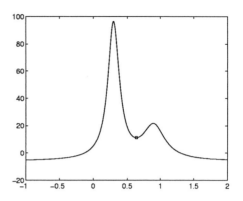

If there does not exist any m-file for the function, or one does not want to litter the directory with small functions, one may create an `inline` object. Here the function `fminbnd` is used, which finds the minimum over an interval, rather then close to a point.

Note that we use the negative definition of the function to find the maximum using `fminbnd`.

```
fneg = inline('-humps(x)');
xs = fminbnd(fneg,-1,2)
xs =
    0.3004
fs = feval(f,xs);
plot(xs,fs,'s')
```

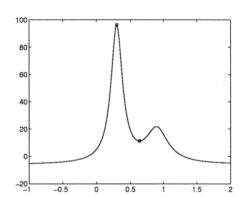

Exercise 52
Inspect the information given by `help eval` *and modify the function* `fdiff` *on page 58 by replacing the two* `feval` *calls by a single call to* `eval`.

Exercise 53
Find the maximum and zeros of the function

$$f(x) = \frac{1}{(x - 0.3)^2 + 0.01} + \frac{1}{(x - 0.9)^2 + 0.04} - 6$$

using `fminbnd` *and* `fzero`.

Exercise 54
Design your own numerical integration function using the approximation

$$\int_a^b f(x)dx \approx \sum_{k=0}^{N-1} f(a + kT)T$$

where $T = (b-a)/N$. *Syntax:* `I=myint(f,a,b,N)`. *Compute an approximation to*

$$\int_0^1 xe^{x^2}\,dx$$

Exercise 55
Construct a general Newton method for solving nonlinear equations $f(x) = 0$. Approximate the differentiation with a fractional difference,

$$f'(x) \approx \frac{f(x+h) - f(x)}{h}$$

where the parameter h is given as an input parameter to your function. Use the syntax:

`x=findzero(f,x0,tol,h)`

Compare your results with the ones obtained with the MATLAB function `fzero`.

Exercise 56
Extend the example `fdiff` given on page 58 so that it chooses the parameter h automatically given a relative error bound ε on the derivative estimate. Let

$$\hat{f}_1'(x) = \frac{f(x+h) - f(x)}{h} \qquad \hat{f}_2'(x) = \frac{f(x) - f(x-h)}{h}$$

Start with some fixed h and decrease it gradually until the relative difference between the two estimates is less than ε, that is until

$$|\hat{f}_1'(x) - \hat{f}_2'(x)| \leq \varepsilon|\hat{f}_2'(x)|$$

Syntax: `fprime=fdiff(f,x,ep)`.

Part 2

ADVANCED PROGRAMMING

11 Data Structures

MATLAB was originally developed as a high level interface to a set of matrix computation library routines. As mentioned in the preface to the book the acronym MATLAB actually stands for MATrix LABoratory. Therefore, the basic data structure in MATLAB is a two-dimensional matrix containing real or complex floating point numbers. Most functionality in MATLAB is provided for operating on this two-dimensional matrix data structure. Character arrays is another data type available in the MATLAB environment. Character arrays are used for storing text data. However, as discussed in Section 4, the character array is actually stored in MATLAB as a vector of numbers defined by the ASCII-code of the characters in the string.

In this section, we will illustrate some other data structures that extend and complement the matrix structure in different ways. The data structures covered by this section are sparse matrices, multi-dimensional arrays, cell arrays and structs.

11.1 Sparse Matrices

Contents: Sparse Matrices
Functions:
`sparse, spy, full, nnz, find, speye, spones, spdiags, sprand,`
`sprandn, issparse.`

MATLAB can store and operate on very large matrices. The computer memory size and computational performance will determine the maximum size of problem that is possible to study. However, in several engineering applications there are problems defined by very large matrices having just a few non-zero elements. For this purpose there is a special data structure in MATLAB devoted to computations performed on such sparse matrices.

A guided tour 28 (Sparse Matrices)

The general calling syntax for creating a sparse matrix is **sparse(i,j,s)** where i and j are integer vectors of equal length listing the matrix coordinates (row and column) where the sparse matrix is non-zero. The vector **s** contains the values of the matrix in the matrix entries specified by i and j.

The function **spy** plots the "sparsity pattern" of a sparse matrix.

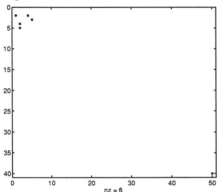

The matrix S is indeed very sparse, having only 0.3% non-zero elements, as is revealed by the **nnz**-command.

A sparse matrix can be converted into a regular matrix by using the command **full**. Listing the workspace variables reveals that MATLAB indeed saves a severe amount of memory space when declaring a matrix to be sparse.

The MATLAB workspace area (available from the 'View' menu in MATLAB 6 and higher) distinguishes sparse and full matrices as shown in the example to the right.

```
>> S = sparse([2 2 3 4 5],...
              [1 4 5 2 2],...
              [10 11 12 13 14])
S =
   (2,1)        10
   (4,2)        13
   (5,2)        14
   (2,4)        11
   (3,5)        12
>> S(40,50) = 15
S =
   (2,1)        10
   (4,2)        13
   (5,2)        14
   (2,4)        11
   (3,5)        12
   (40,50)      15
>> nnz(S)/prod(size(S))
ans =
    0.0030
>> spy(S)
```

```
>> Sfull = full(S);
>> whos
  Name    Size    Bytes   Class

  S       40x50     384   sparse array
  Sfull   40x50   16000   double array
>> workspace
```

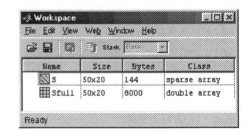

All arithmetic, logical and indexing operations can be applied to sparse matrices, or to mixtures of sparse and full matrices. Operations on sparse matrices return sparse matrices and operations on full matrices return full matrices. In most cases, operations on mixtures of sparse and full matrices return full matrices.

Note that the operation .* on a mixture of sparse and full matrices will yield a sparse matrix, while the addition operator will not. The logical test == yields a sparse matrix with no non-zero entries. Note that the matrix S4 therefore will require more storing space than a corresponding full matrix.

We have seen that the requirement on memory decreases using sparse matrices. All arithmetic operations on sparse matrices utilize the sparsity pattern in such a way that the zero-entries are ignored whenever possible. Besides arithmetics, many algebraic functions are also over-loaded to the sparse data format. In this way, the computation time is substantially decreased in many cases.

Here, we first construct a *Toeplitz* matrix, whose diagonal structure is highlighted by the spy plot, and then the eigenvalues are computed. The example shows that computation time is less for the sparse matrix.

Figure 27.7 in Appendix E shows another example on how to visualize a large matrix with a particular structure.

```
>> S1 = S*S'; issparse(S1)
ans =
     1
>> S2 = Sfull.*S; issparse(S2)
ans =
     1
>> S3 = Sfull+S; issparse(S3)
ans =
     0
>> S4 = Sfull == S; issparse(S4)
ans =
     1
>> S5 = sin(S); issparse(S5)
ans =
     1
>> Xfull=toeplitz([1 0 2 0 3...
                   zeros(1,50)]);
>> X=sparse(Xfull);
>> spy(X)
>> tic,eig(Xfull);toc
elapsed_time =
    0.0161
>> tic,eig(X);toc
elapsed_time =
    0.0018
```

11.2 Multidimensional Arrays and Cell Arrays

Contents: Multidimensional Arrays and Cell Arrays.
Functions:
`repmat, cat, ndims, permute, ipermute, squeeze,`
`varargin, varargout.`

A guided tour 29 (Multi-dimensional arrays)

Sparse matrices are always two-dimensional, but MATLAB can actually handle full matrices of any dimension. The example to the right shows how to extend a regular two-dimensional matrix into a three-dimensional one.

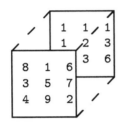

The generalization of matrix transpose `.'` to higher dimension is called **permute**. With this function, the dimensions can be reorganized in any order.

The transpose of a two-dimensional matrix M can be written **permute(M,[2 1])**. In two dimensions, taking the transpose twice yields the untransposed matrix as a result. The more complex permutation in higher dimension provided by **permute** does not have this property in general. MATLAB provides a function **ipermute** that computes the inverse permutation of a matrix

```
>> A(:,:,1)=magic(3)
A =
     8     1     6
     3     5     7
     4     9     2
>> A(:,:,2)=pascal(3)
A(:,:,1) =
     8     1     6
     3     5     7
     4     9     2
A(:,:,2) =
     1     1     1
     1     2     3
     1     3     6
```

```
>> permute(A,[3 1 2])
ans(:,:,1) =
     8     3     4
     1     1     1
ans(:,:,2) =
     1     5     9
     1     2     3
ans(:,:,3) =
     6     7     2
     1     3     6
>> permute(magic(2),[2 1])
ans =
     1     4
     3     2
```

The content of a three-dimensional array can be illustrated in a 3D-plot. The permutation of the matrix A then corresponds to a change of coordinate system from (x, y, z) to (z, x, y). In the plot below, this can be interpreted as a change of camera angle.

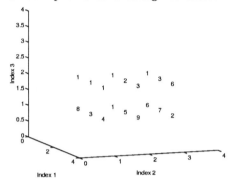

Multidimensional arrays can also be constructed from commands like **randn**, **ones**, and **zeros**.

The function **repmat** replicates a matrix along any dimension. In the example to the right, a scalar matrix with an infinite entry is replicated twice in the first and third dimension and three times in the second dimension.

Multi-dimensional matrices can also be constructed by concatenating matrices of compatible dimensions using the MATLAB function **cat**.

The function **ndims** determines the highest dimension of the matrix. Singleton dimensions can sometimes be inserted after a MATLAB operation on a high dimensional matrix. These dimensions are removed using the command **squeeze**.

```
>> axis([0 4 0 4 0 4]), hold on
>> for i=1:size(A,1)
  for j=1:size(A,2)
   for k=1:size(A,3)
     text(i,j,k,num2str(A(i,j,k)))
   end
  end
end
>> xlabel('Index 1')
>> ylabel('Index 2')
>> zlabel('Index 3')
>> view([70 10])
```

```
>> B = randn(4,3,2,3);
>> C = repmat(Inf,[2,3,2])
C(:,:,1) =
    Inf    Inf    Inf
    Inf    Inf    Inf
C(:,:,2) =
    Inf    Inf    Inf
    Inf    Inf    Inf

>> D = cat(4,magic(2),pascal(2))
D(:,:,1,1) =
    1    3
    4    2
D(:,:,1,2) =
    1    1
    1    2
>> ndims(D)
ans =
    4
>> D = squeeze(D)
D(:,:,1) =
    1    3
    4    2
D(:,:,2) =
    1    1
    1    2
>> ndims(D)
ans =
    3
```

A guided tour 30 (Cell arrays)

The cell array is a data structure that can contain any other data structure, even another cell array. Cell arrays can be assigned by wrapping the right hand side of the expression in curly brackets, {}. Curly brackets are also used for indexing cell arrays.

'text'	1 0 0 1
i	{'A',pi}

Another way to assign cell arrays is to index the left hand side with curly brackets. Note that unassigned entries in the cell array are filled with empty matrices, [].

Indexing a cell array with a colon operator yields what is called a "comma separated list" of the entries in the cell array. This is the equivalent to typing each of the entries of the cell array in the MATLAB window, separated by commas.

The comma separated list is the key to building functions taking any number of input parameters. The function to the right utilizes this for computing the average score of an exam taking any number of students. The input to the function is assigned the cell array **varargin** and the output is assigned to a corresponding cell array **varargout** so that both input and output can be of varying number.

```
>> A = {'text',eye(2);i,{'A',pi}}
A =
    'text'          [2x2 double]
    [0+ 1.0000i]    {1x2 cell  }
>> A{1,1}
ans =
text
>> A{2,2}
ans =
    'A'      [3.1416]
```

```
>> Cl{1,2} = 'Mr. Magoo'
Cl =
    []      'Mr. Magoo'
```

```
>> A{:}
ans =
text
ans =
        0 + 1.0000i
ans =
    1       0
    0       1
ans =
    'A'      [3.1416]
```

File name: exam.m
```
function varargout=exam(varargin)

names = {varargin{1:2:end}};
points =[varargin{2:2:end}];
av = mean(points);
[sp,ind] = sort(points);
names = {names{fliplr(ind)}};
if nargout < 1
   disp(['Average grade: ',...
            num2str(av)])
else
   varargout{1} = av;
   varargout(2:nargout) = ...
            {names{1:nargout-1}};
end
```

The input is an alternating list of students with name and grade on the exam. If there is a variable assigned to the output from the function, the output will consist of the average grade over the student population followed by the name of the best students in ascending order. If there is no output assigned, the function will display a message presenting the average grade.

Like regular arrays, cell arrays can also be of arbitrary dimension. The example to the right shows how to extend the previously defined two-dimensional cell array A from Guided Tour 29 on page 64 into a three dimensional cell array.

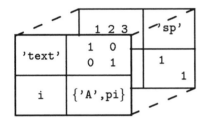

```
>> [av,n1,n2]=exam('Bob',2,...
                   'Fred',5,'Liv',1)
av =
    2.6667
n1 =
Fred
n2 =
Bob
>> exam('Nic',2,'Fred',5,'Lena',1)
Average grade: 2.6667
>> B={1:3,'sp';[],speye(2)}
B =
    [1x3 double]    'sp'
             []    [2x2 sparse]
>> A = cat(3,A,B)
A(:,:,1) =
    'text'          [2x2 double]
    [0+ 1.0000i]    {1x2 cell  }
A(:,:,2) =
    [1x3 double]    'sp'
             []    [2x2 sparse]
```

11.3   Structs

The data structure **struct** is a clustering of different data types under one name. We use structs whenever we are handling data objects described by strings, vectors and matrices of different sizes, but we still want to have just one variable name for the complete object. In struct arrays, each field has a logical name that is very useful for giving the field a natural description.

A guided tour 31 (The struct datatype)
In this example, we design an object for a recursion, where the Fibonacci series in exercise 48 is one example. That is, the recursion is defined by the coefficients p_i, represented by a vector. Each recursion is identified by a name tag. Two different ways to define **r** are presented to the right.

```
>> % Alternative 1
>> r=struct('Poly',[1 1],...
    'Name','Fibonacci');
>> % Alternative 2
>> r.Poly=[1 1];
>> r.Name='Fibonacci';
```

A recursion can be defined as

$$y[k] = p_1 y[k-1] + p_2 y[k-2] + ... + p_n y[k-n].$$

Generally, to simulate a polynomial of degree n, we need to define the first n values in the series. These are the initial conditions. Here we define $y[k] = 0$, $k = 1, 2, ..., n-1$, and $y[n] = 1$, and start the simulation at index $n+1$. The recursion is here computed for N=10 time steps.

```
r:
```

r.Poly	[1 1]
r.Name	'Fibonacci

We can add another recursion, whose output $y[k]$ resembles the movement of a damped spring, like a car moving too quickly over a speed hump.

A mathematical recursion for such a movement is

$$y[k] = 0.9y[k-1] - 0.81y[k-2]. \quad (11.1)$$

We can then address all name fields in a simple way, or check the first coefficient in the first object. For data base applications, a search for specific features are easily implemented.

r:	r(1)	r(2)

r.Poly	[1 1]	[0.9 -0.81]
r.Name	'Fibonacci	'Spring'

```
>> n=length(r.Poly);
>> N=10;
>> y=zeros(N,1);
>> y(n)=1;
>> for i=n+1:N
     y(i)=r.Poly*y(i-1:-1:i-n);
end
>> plot(y,'.-')
```

```
>> r(2).Poly=[0.9 -0.81];
>> r(2).Name='Spring';
>> r
r =
1x2 struct array with fields:
    Poly
    Name
>> r(1).Poly(1)
ans =
    1
>> r(:).Name
ans =
Fibonacci
ans =
Spring
```

In the next example, we illustrate a mixture of vector index, structures and arrays in a larger database application. The considered database is a tool for recording music, sorting the information and playing the tunes in different order.

A guided tour 32 (A database example, CD collection)

The first thing to decide on is the structure of the database, and what a suitable data format is. In the example, we use a vector `db(i)`, where each element is a struct. The struct contains fields for artist, album and number of songs. For each song, the struct contains one vector for the length, one character array for the title and one array with vectors containing the sampled music.

The function `cddbinput` is a small but complete interactive recording tool for inputting data to the struct, and using Windows drivers for the microphone (line in) input. By attaching a CD player to the line input on a PC, and some time consuming keyboard inputting and timing, a database is built up. We assume here that a subset of three albums are recorded.

Example:

File name: `cddbinput.m`

```
function db=cddbinput(db)

n=length(db)+1;
db(n).Artist=input('Artist','s');
db(n).Album=input('Album','s');
m=input('Nr of Songs');
db(n).NrSongs=m;
fs=8198;
for i=1:m
  T=input('Length of next song');
  db(n).Song{i}=input('Name','s');
  disp('Press a key to start')
  pause
  y=wavrecord(fs*T,fs);
  db(n).Data(i)=y;
  db(n).Length(i)=T;
end
```

```
>> db=cddbinput(db)
db =
1x3 struct array with fields:
    Artist
    Album
    NrSongs
    Length
    Songs
    Data
```

The function **sortrows** is powerful for sorting character matrices in *alphabetical order*. The function **cddbsort** sorts the database either with respect to the artist, or the album or the songs. There are many conversions from character arrays to strings and character matrices using **str2mat**. The middle column of the output matrix **char(32*ones(n,3))** creates a matrix of size $n \times 3$ with blank spaces.

Example:

```
>> cddbsort(db,'Artist')
ans =
Abba      Gold
Abba      Silver
ZZ Top    Bronze

>> cddbsort(db,'Album')
ans =
Bronze    ZZ Top
Gold      Abba
Silver    Abba

>> cddbsort(db,'Songs')
Another one
Deliminator
My favorite
Song 1
Song 2
Yet one more
```

File name: `cddbsort.m`

```
function cddbsort(db,arg)

n=length(db);
switch arg
 case 'Artist'
  list=str2mat(db.Artist);
  [a,ind]=sortrows(list);
  [list(ind,:),...
    char(32*ones(n,3)),...
    str2mat(db(ind).Album)]
 case 'Album'
  list=str2mat(db.Album);
  [a,ind]=sortrows(list);
  [list(ind,:),...
    char(32*ones(n,3)),...
    str2mat(db(ind).Album)]
 case 'Songs'
  list=[];
  for i=1:n;
   list=str2mat(list,...
      str2mat(db(i).Songs));
  end
  disp(sortrows(list))
end
```

To play the music, we want the usual functions of choosing random or chronological playing order. Unlike CD changers, we can here also choose to play all records with a particular artist by searching the database for a certain string in the artist field. Before sending the music data to Windows drivers, information of the current song is displayed.
Example:

```
>> cddbplay(db,'Abba','Order')
Song 1 by Abba from Gold
Song 2 by Abba from Gold
My favorite by Abba from Silver
Another one by Abba from Silver
Yet one more by Abba from Silver

>> cddbplay(db,'All','Random')
Song 1 by Abba from Gold
Song 1 by Abba from Gold
My favorite by Abba from Silver
Deliminator by ZZ Top from Bronze
Song 2 by Abba from Gold
Another one by Abba from Silver
Song 1 by Abba from Gold
Another one by Abba from Silver
Song 1 by Abba from Gold
Yet one more by Abba from Silver
```

With a graphical user interface, this database tool might be close to realistic. The most obvious drawback with the demonstrated data structure is that the musical data should not be stored in the same structure as the other data. More logical is to save each song on a separate file, and then replace the db(i).Data{j} vector with the file name. The modifications of this code would be quite small.

File name: `cddbplay.m`

```
function cddbplay(db,Name,How)

n=length(db);
ind=[];
for i=1:n
  if strcmp(db(i).Artist,Name)|...
     strcmp('All',Name)
    ind=[ind i];
    N(i)=db(i).NrSongs;
  end
end
n=length(ind);
switch How
  case 'Order'
    for i=1:n;
      for j=1:N(i);
        disp([char(db(i).Songs(j)),...
        ' by ',char(db(i).Artist),...
        ' from ',...
        char(db(i).Album)])
        soundsc(db(i).Dataj)
      end
    end
  case 'Random'
    for k=1:10;
      i=ceil(n*rand(1));
      j=ceil(N(i)*rand(1));
      disp([char(db(i).Songs(j)),...
      ' by ',char(db(i).Artist),...
      ' from ',...
      char(db(i).Album)])
      soundsc(db(i).Dataj)
    end
end
```

12   Object Orientation

The main difference between a MATLAB `object` and a `struct` is the possibility to *overload* functions to an object. That means that there may be several functions with the same name, and the object type determines which one is used. For instance, assume that we have defined an object referenced to the variable r. We can design our own plot function `plot.m`, and actually execute it with `plot(r)`. MATLAB detects that r is an object of the considered type, and automatically switches to our plot function. By redefining r to a vector of real numbers, the default plot-function will automatically be used. Suppose we want to design an object called `rec`. The requirement are that:

- There exists a directory called @rec in one of the directories in the MATLAB search path (the directory @rec itself does not need to be included in the search path, MATLAB checks if there are any directories with a name starting with @).

- There is a file `rec.m` in the directory @rec.

- All functions in the directory @rec are overloaded to the object `rec`.

A guided tour 33 (Objects and classes)
The m-file `rec.m` defines the fields of an object `rec` as a struct. To define the content of the object in the MATLAB window, just give an assignment r=rec({Value1 Value2..}), where you list all field values. For example, the Fibonacci and spring recursions are defined as:

```
>> r1=rec({[1 1] 'Fibonacci'});
>> r2=rec({[0.9 -0.81] 'Spring'});
```

File name: @rec/rec.m
```
function rout=rec(rin)

if nargin==0;
    % Create an empty struct
    rout=struct('Poly',[],...
    'Name','Unnamed');
    rout=class(rout,'rec');
elseif isa(rin,'rec')
    % is rin the object rec?
    rout=rin;
    % Then just pass it out
elseif size(rin)==[1 2]
    rout.Poly=rin1;
    rout.Name=rin2;
    rout=class(rout,'rec');
else
    error([num2str(rin),...
    ' is not an object ''rec'''])
end
```

The Fibonacci series is an example of a divergent series, while the spring model is convergent and eventually comes to rest with $y[k] = 0$. Theoretically, divergence or convergence can be assessed by the roots of the characteristic polynomial, and we overload this computation on `roots`.

File name: @rec/roots.m
```
function lam=roots(r);

lam=roots([1 -r.Poly]);
```

It may be instructive to present the recursion in a text based format using a function we may associate with `disp`. The example here first generates a string describing the recursion, then computes $y[k]$ for $k = 1, 2, ..., N$, and displays the result in a table. Example:

```
>> disp(r1)
Recursion: y(k) = 1y(t-1) + 1y(k-2)
-------------
Iter. | Value
-------------
   1 | 0
   2 | 1
   3 | 1
   4 | 2
   5 | 3
   6 | 5
   7 | 8
   8 | 13
   9 | 21
-------------
```

An even more elegant solution is to use the overloaded `display` function. This function is called whenever the object name is evaluated without a trailing semi-colon in the workspace:

```
>> r1
```

File name: `@rec/disp.m`

```
function t=disp(r,N)

t=['y(k) = ',...
    num2str(r.Poly(1)),'y(t-1)'];
for i=2:length(r.Poly);
    t=[t,' + ',...
        num2str(r.Poly(i)),...
        'y(k-',num2str(i),')'];
end
if nargout==0;
    n=length(r.Poly);
    y=zeros(N,1);
    y(n)=1;
    for i=n+1:N
        y(i)=r.Poly*y(i-1:-1:i-n);
    end
    disp(['Recursion: ',t])
    disp('-------------')
    disp('Iter. | Value')
    disp('-------------')
    for i=1:N
        disp(['    ',num2str(i),...
            ' | ',num2str(y(i))])
    end
    disp('-------------')
end
```

File name: `@rec/display.m`

```
function display(r)

disp(r,9)
```

thus gives the same answer as above.

We will next construct a plot function for this object, and re-define the meaning of + (plus) for two objects in this class. To sum up, we then have created a directory that should look as follows:

```
>> dir ../mfiles/@rec
.            disp.m      plot.m     rec.m
..           display.m   plus.m     roots.m
```

To illustrate the recursions, we can write a dedicated plot function which plots $y[k]$ for $k = 1, 2, ..., N$ and also the roots of the characteristic polynomial.

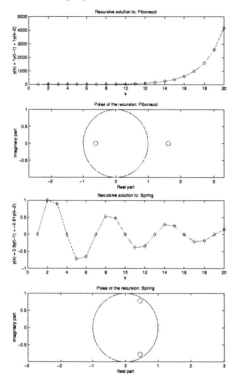

We can redefine the + operator to add recursions, which here is done by using **conv** on the polynomials. The new object is also assigned an appropriate name.
Example:

```
plot(r1+r2)
```

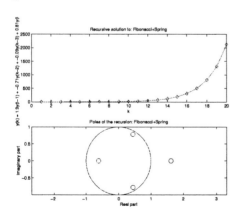

File name: **@rec/plot.m**

```
function plot(r,N)


n=length(r.Poly);
y=zeros(N,1);
y(n)=1;
for i=n+1:N
    y(i)=r.Poly*y(i-1:-1:i-n);
end

figure
subplot(2,1,1)
plot(1:N,y,'-r',1:N,y,'db')
title(['Recursion: ',r.Name])
t=disp(r);
ylabel(t),xlabel('k')
subplot(2,1,2)
% axis([-1 1 -1 2])
phi=0:0.1:2*pi;
plot(sin(phi),cos(phi))
hold on
lam=roots(r);
scatter(real(lam),imag(lam),100)
axis('equal')
title(['Poles of recursion: ',...
        r.Name])
xlabel('Real part'),
ylabel('Imaginary part')
```

```
plot(r1,20)
plot(r2,20)
```

File name: **@rec/plus.m**

```
function r=plus(r1,r2)

r=rec;
p1=r1.Poly;
p2=r2.Poly;
p=conv([1 -p1],[1 -p2]);
r.Poly=-p(2:end);
r.Name=[r1.Name,'+',r2.Name];
```

MATLAB supports several other concepts of object orientation. Encapsulation of data is possible by defining data members and only allowing them to be altered by special member methods. The methods themselves are encapsulated by defining functions stored in a subdirectory called `private`. New classes can be defined through multiple or single inheritage and you can create new classes of objects by aggregation, letting a new class be defined by combination of other classes. See the on-line documentation for more information and examples.

13 Graphical Object Orientation and User Interfaces

13.1 Graphical objects

All graphics are represented as *graphical objects*, each labeled by a *handle*. These contain a number of properties having arguments of different data types. Generally, the properties are accessed by **set** and **get**. Do **get(handle)** to print out a list of the different properties and their current values on the screen. A similar listing is obtained with **set(handle)**. The difference is that only properties belonging to an enumerable list (like the properties **'on'** and **'off'**) are shown, and the current setting is highlighted.

The hierarchical structure of graphical objects is listed below:

0 System parameters are kept in the root object, always labeled with handle 0.

1 The children of the system are the figure windows, labeled with integer handles. **gcf** (get current figure) returns the integer handle of the current figure (the last generated one or the last one marked with the mouse).

2 The children of each figure are **axes**, **menu**, **uicontrol**. These are labeled with handles of real numbers, where the integer part is the figure window they are located in. **gca** (get current axis) returns the handle of the current axis in the current figure and **gco** (get current object) returns the handle of the current object in the current figure. In both cases, current means the last one marked with the mouse, or if this does not apply, the last generated one.

3 The children of an **axes** object are lines, markers, images **etc**.

An example of fields in the root is given below (there are many more fields, but these should give an idea of what type of information is kept here):

```
get(0)
        CommandWindowSize = [80 38]
        CurrentFigure = [7]
        Diary = off
        DiaryFile = diary
        Echo = off
        ErrorMessage = [ (1 by 152) char array]
        FixedWidthFontName = Courier
        Format = short
        FormatSpacing = compact
        Language = iso_8859_1
        More = off
        PointerLocation = [279 122]
        ScreenSize = [1 1 1280 1024]
        Units = pixels

        Children = [ (7 by 1) double array]
        Tag =
        Type = root
```

13.2 Default settings

There are an extensive list of default values. The factory settings can be listed as follows (again, only a few properties are listed here, there are several hundreds of them):

```
get(0,'factory')
ans =
```

```
              factoryFigurePosition: [100 100 660 520]
                  factoryTextColor: [1 1 1]
              factoryTextFontName: 'Helvetica'
              factoryTextFontSize: 10
             factoryTextFontUnits: 'points'
             factoryLineLineStyle: '-'
             factoryLineLineWidth: 0.5000
                factoryLineMarker: 'none'
```

You can change any of these. For instance, to change the place where new figure windows appear on the screen to the upper right corner, type something like

```
s=get(0,'ScreenSize');
set(0,'DefaultFigurePosition',[s(3)-400 s(4)-300-70 400 300])
```

Here 70 pixels are reserved to the figure name header. Just change **factory** in the list to **Default**, when setting new properties.

For some reason, the default values in the figure windows are set too small for most publication purposes. There are three alternatives to change this:

- Change the properties after each plot has been created, by saving the handles of each object.

- Change the default values of all objects created within a figure.

- Change the default values of all objects in all figure windows.

The examples below illustrate these principles.

A guided tour 34 (Changing the default values)

The function **plotfix** changes the default settings of important figure properties so that the lines and text will be clearly readable in publications, see also Section 27.

File name: `plotfix.m`

```
function plotfix
% Change the default font size
% and line width in
% current figure.
% Replace gcf with 0 to make
% it global to all new figures.
set(gcf,...
  'DefaultAxesFontSize',16,...
  'DefaultLineLineWidth',1,...
  'DefaultAxesLineWidth',1,...
  'DefaultLineMarkerSize',10,...
  'DefaultAxesGridLineStyle',...
        '--',...
  'DefaultTextFontSize',14);
```

The alternative to changing the properties *before* the objects are created, is to do it afterwards. Here you have two options:

- Save the object handles (when an output argument is specified, each graphical function returns the handle), and then use **set**.

- Change the properties by the GUI **propedit**. This can be done to saved figures. You can either

  - use the **.fig** format which can later be opened by **open**. To save in **.fig**, either use **saveas** (as an example, **saveas(fignr,filename,'fig')**) or choose **Save** in the **File** menu of the figure to save it with extension **.fig**. The figure can be reloaded from the same menu, for example during another MATLAB session.

  - or generate an m-file with **saveas(fignr,filename,'m')**, and the figure is re-created when calling the m-file.

Note that figures saved with **print** cannot be altered afterwards in MATLAB.

See also the comments in Section 27, where the recommendation is to save the code used to generate the figure.

Another option is to change the default values for the complete session, by setting this value on the root level in the file **startup.m**.

File name: **startup.m**

```
function startup
% Change the default font size
% and line width forever
set(0,...
  'DefaultAxesFontSize',14,
  'DefaultLineLineWidth',1,
  'DefaultAxesLineWidth',1,
  'DefaultAxesGridLineStyle','--'
  'DefaultTextFontSize',14);
```

13.3   Graphical User Interface (GUI)

We will now show an example using most of the features of the graphical objects. The application is to illustrate a *chirp signal*, which is defined as a sinusoid with increasing frequency. Mathematically, a chirp sweeping linearly from f_0 Hertz to f_1 Hertz computed in N points separated T_s seconds can be defined as

$$y[k] = \sin\left(2\pi f_0 k T_s + \frac{k^2(f_1 - f_0)T_s}{2N}\right), \quad k = 1, 2, \ldots, N. \quad (13.2)$$

The local frequency of this sinusoid can be defined as $f_0 + (f_1 - f_0)k/N$ Hertz (this is how the ear interprets the sound of this signal). The MATLAB function **chirp** computes this, and the sweeping can be changed to quadratic or logarithmic.

The goal is to develop a GUI as shown in Figure 13.1 for examining chirp signals with the following specifications:

1. Show the spectrogram (compare to section 25).

2. Play it as sound.

3. To be able to change the start and end frequency of the chirp as edit fields.

4. To be able to change the simulation time with a slider.

5. To switch between the different frequency sweep methods in `chirp`.

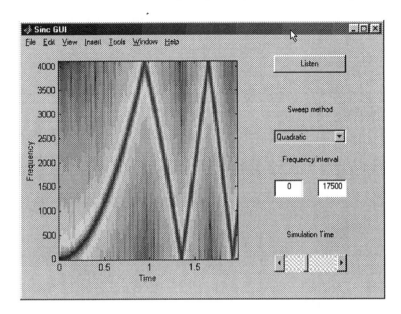

Figure 13.1 The GUI described in Guided Tour 35, saved as a screen dump.

The GUI will be generated by writing a file with handle graphics, and the alternative to this would be to use **guide**, as shortly discussed after the guided tour. The objects are hierarchically structured as shown in Figure 13.2.

A guided tour 35 (Graphical user interfaces)

First, a new figure window is created. It is designed in centimeters, but can also be chosen in pixels and related to the *screen size*. To get the resolution, type s=get(0,'Screensize'); The advantage with centimeters is that it is easier to set the paper size, so a hardcopy printout keeps the aspect ratio. The function **print** only works for **axes** objects in figures, so use the printing option in the **file** menu. It is a good tip to change the units to 'Normalized' afterwards, so that everything scales nicely when the window size is changed with the mouse. The last line defines an **axes**, where the spectrogram will appear.

```
hf=figure('Name','Sinc GUI',...
  'NumberTitle','off',...
  'Units','Centimeters',...
  'Position',[1 3 15 10],...
  'Units','Normalized',...
  'PaperUnits','Centimeters',...
  'PaperPosition',[0.25 3 18 12]);
ha=axes('Position',...
      [0.1 0.15 0.5 0.8]);
```

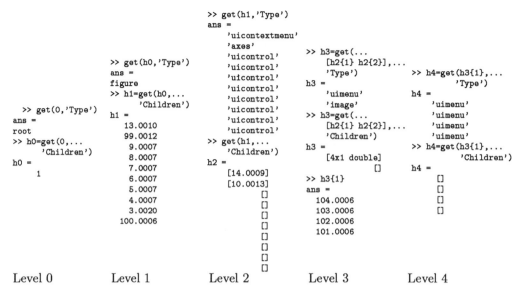

Figure 13.2 Hierarchical organization of the graphical objects in Figure 13.1. The child of the root is here just one figure window, and its children are one uicontextmeny, one axes (whose child is an image) and eight uicontrol.

To the right of the spectrogram, we want a column with controls. The first one is a push button. When it is pressed, a field called *call-back* is set to `listen`. The call-back describes the action that is executed after pushing the button. Pushing the button will be exactly the same as typing `listen` in the MATLAB window. The consequence of this is that it prevents the use of local variables in the call-back (`listen(y)` does not work for instance), and it also prevents the use of sub-functions (`listen` is defined at the end of the GUI function).

```
hui1=uicontrol(hf,...
  'Style','PushButton',...
  'String','Listen',...
  'Units','normalized',...
  'Position',[0.7 0.9 0.2 0.07],...
  'Callback','listen');
```

The next `uicontrol` just defines text as a title of the next control.

```
uicontrol(hf,...
  'Style','Text',...
  'String','Sweep method',...
  'Units','normalized',...
  'Position',[0.7 0.7 0.2 0.07]);
```

The *pop-up menus* are defined by a list of options in the `String` field, separated by bars. The user chosen option is read off as the field `Value` (an integer), which is here as a default value set to the second option (quadratic).

The next `uicontrol` is another text string.

An *edit field* in the GUI contains its values as a `String`. We have to remember to convert the string to a numerical value when it is read off. It is customary to have a white background color in edit fields. The call-back is here executed when pressing **return**.

The second edit field is defined analogously.

The *slider* first gets a title.

The slider has the extra fields `Min` and `Max`, and `Value` sets the default value. Each time the user moves the slider with the mouse, `Value` changes and the `simandplot` function is called.

```
hui2=uicontrol(hf,...
  'Style','PopUp',...
  'String',...
  'Linear|Quadratic|Logarithmic',...
  'Value',2,...
  'Units','normalized',...
  'Position',[0.7 0.6 0.2 0.07],...
  'Callback','simandplot');
hui3=uicontrol(hf,...
  'Style','Text',...
  'String','Frequency interval',...
  'Units','normalized',...
  'Position',[0.7 0.5 0.2 0.07]);
hui4=uicontrol(hf,...
  'Style','Edit',...
  'String','0',...
  'Callback','simandplot',...
  'BackgroundColor','w',...
  'Units','normalized',...
  'Position',[0.7 0.4 0.08 0.07]);
hui5=uicontrol(hf,...
  'Style','Edit',...
  'String','17500',...
  'Callback','simandplot',...
  'BackgroundColor','w',...
  'Units','normalized',...
  'Position',[0.82 0.4 0.08 0.07]);
uicontrol(hf,...
  'Style','Text',...
  'String','Simulation Time',...
  'Units','normalized',...
  'Position',[0.7 0.2 0.2 0.07]);
hui6=uicontrol(hf,...
  'Style','Slider',...
  'Min',0,'Max',5,...
  'Value',2,...
  'Callback','simandplot',...
  'Units','normalized',...
  'Position',[0.7 0.1 0.2 0.07]);
```

One tricky part with GUI programming is how to pass data to the call-back functions. As usual, global variables should be avoided. The recommended solution is to use the UserData fields available for all MATLAB objects. If many variables and strings are needed, a struct or MATLAB object should be used, see Section 11.3 or Section 12. When all graphical objects are created, we call the main function simandplot to get the default spectrogram in the axes object.

To read the content of uicontrol and to access axes objects, their handles are needed. There are basically two ways to transport the handles to the call-back functions:

- Create a vector with handles, and use the UserData field of one of the top objects.

- The usually recommended solution is to set the Tag of each handle, and then search for the tag by using findobj. A typical procedure is to define set(h,'Tag','axes1') and when the handle is needed, use h=findobj('Tag','axes1');

Here the findobj is used to find the figure window, whose UserData contains a vector of handles.

Then the pop-up menu, edit field and slider are read off using get, and the sinc function is called. The signal is stored in the UserData field of the pushbutton, where it will be looked for when the push button Listen is pressed and the function listen is called.

Finally, we attach a context menu to the image, where the user can choose between a number of pre-defined color maps in MATLAB. In this way, the impression of the spectrogram image changes considerably.

Note that the uicontextmenu is a child of the figure, and linked to the image by using the field UIContextMenu in the image.

```
set(gcf,'UserData',...
  [ha hui1 hui2 hui3...
  hui4 hui5 hui6])
simandplot
```

File name: simandplot.m

```
function simandplot
% Callback from sincgui
hf=findobj(0,'Name','Sinc GUI');
h=get(hf,'UserData');
switch get(h(3),'Value')
case 1
  method='linear';
case 2
  method='quadratic';
case 3
  method='logarithmic';
end
f0=str2num(get(h(5),'String'));
f1=str2num(get(h(6),'String'));
T=get(h(7),'Value');
y=chirp(0:1/8192:T,f0,T,...
        f1,method);
set(h(2),'UserData',y)
axes(h(1))
specgram(y,[],8192);
```

File name: simandplot (cont'd)

```
hi=findobj(hf,'Type','Image');
hmenu=uicontextmenu;
set(hi,'UIContextMenu',hmenu)
hmenu1=uimenu(hmenu,...
  'Label','Colormap');
```

To the basic menu, we add a sub-menu for different color-maps. Basically, the *color-map* is a 256×3 matrix `colormap`, which assigns an *rgb* value (that is, a color) to each integer between 0 and 255. The function that generated the graphics thus has to compute a suitable grid with 256 bins for all values in the vector/matrix that is illustrated.

There are many pre-defined color maps. Type `help graph3d` to list them.

File name: `simandplot.m` (cont'd)

```
uimenu(hmenu1,'Label','jet',...
  'CallBack',...
  'colormap(''jet'')');
uimenu(hmenu1,'Label',...
  'copper',...
  'CallBack',...
  'colormap(''copper'')');
uimenu(hmenu1,'Label',...
  'gray',...
  'CallBack',...
  'colormap(''gray'')');
```

The `listen` function reads the sinc signal y, and uses the `sound` facility.

PC users may prefer to use `wavplay`, that uses Windows drivers. In any case, stereo sound is supported by MATLAB, but this also depends on your sound card.

File name: `listen.m`

```
function listen
% Callback from sincgui
hf=findobj(0,'Name','Sinc GUI');
h=get(hf,'UserData');
y=get(h(2),'UserData');
sound(y)
```

13.4 Constructing a GUI using `guide`

The object oriented graphics is powerful for generating nice abstract interfaces to your m-files. But admittedly there is a certain threshold for getting started with coding the graphical GUI. A very powerful shortcut to create a GUI is to use another GUI called `guide` to specify the handle graphics, and align buttons and graphics in a quick way. Just run the command `guide` and use the palette on the left hand side of the `guide`-window to edit it so that it look like the window shown in the figure below

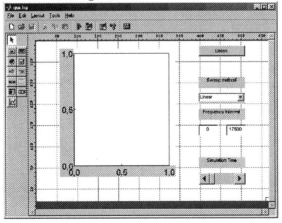

In the `guide`-window you can save the created GUI as a `fig`-file, defining the layout and an m-file, for defining the call-backs. The automatically generated m-file contains information on how to edit and alter the call-back logic for the GUI.

14 Optimizing MATLAB Code

The programming environment in MATLAB is quite convenient for producing small segments of code from scratch quite fast. There is no need to compile the source code, there is no need to declare variables before they are used and there are no complicated library dependencies that need to be resolved before running and testing the developed program. Since MATLAB code is interpreted at run-time it is very easy to alter, adjust and test new implementation ideas as they appear, without the need to recompile and build a new executable version of the developed program.

This makes development in MATLAB quite fast compared to for example development in C or any other compiled language. One potential disadvantage with MATLAB is that the performance of the developed code sometimes is far from the performance one can achieve using a compiled language. However, the sum of development time and execution time is often very competitive for smaller problems and for rapid prototyping.

Nevertheless, MATLAB code can sometimes run quite slow, at least compared to compiled C-code. This is generally the fact since MATLAB is an interpreted system and not compiled. However, there are several ways to improve the execution time of MATLAB code. In order to write efficient code in MATLAB, one has to understand some of the underlying techniques MATLAB utilizes for memory allocation and for program execution. In this section, we cover the following topics for code optimization.

- Vectorization to eliminate for-loops

- Pre-allocation of large variables in memory

- Saving memory space by using compact data types

- Function call by reference and call by value

- Finding the bottleneck in the programs with **profile**

All arithmetic and logical operations in MATLAB can operate along any dimension of a multi-dimensional array using just one function call. In compiled languages, like C, it is often convenient and efficient to write small for-loops for performing operations on data arrays. These for-loops are hidden inside the arithmetic functions provided by the MATLAB kernel. Therefore it is much more efficient to write code that operate on an entire array in a vectorized manner.

A guided tour 36 (Code optimization)

A famous algorithm for generating independent uniform random variables $\{u_i\}_{i=1}^N$ automatically sorted in ascending order is given by

1. Generate $\tilde{u}_i$ for $i = 1, \ldots, N$

2. Set $u_N = \sqrt[N]{\tilde{u}_N}$

3. And $u_i = u_{i+1}\sqrt[i]{\tilde{u}_i}$ for $i = N-1, \ldots, 1$

This algorithm has complexity $O(N)$. The alternative is to sort N uniform samples, the best algorithms for this task are $O(N \log(N))$.

File name: **rands1.m**
```
function u = rands1(n)
utilde = rand(1,n);
u(n) = utilde(n)^(1/n);
for i = n-1:-1:1
    u(i) = u(i+1)*utilde(i)^(1/i);
end
```

File name: **rands2.m**
```
function u=rands2(n)
u = fliplr(cumprod(...
    rand(1,n).^(1./(n:-1:1))));
```

Two implementations of this algorithm are shown to the right. The first one is a straight-forward implementation of the algorithm from the pseudo-code above. The second one utilizes the vector notation of MATLAB to generate all samples at once. The function cumprod computes the cumulative product of the entries in a vector and fliplr just flips the whole vector so that the first entry comes last. fliplr is just used so that the output is in ascending instead of descending order. When timing the functions the optimized code is seen to yield a substantial improvement over the direct pseudo-code implementation.

Since there are no variable declarations in MATLAB, the system needs to allocate memory for variables as they appear in the code. The most efficient way to do this is to allocate the variable once, and not to increase the variable size gradually. Sometimes, it can even be more efficient to allocate a too large array and later remove unwanted entries. This is utilized in the MATLAB function primes which is part of the standard MATLAB distribution. An abbreviated version of the primes-code is shown to the right. The output vector p should contain all prime numbers less than or equal to the input parameter n. After a special test for the case n=1, the vector p is allocated as the number 2 followed by all odd numbers up to n. All odd numbers will not be prime, but the prime number we are interested in finding are among them.

We know that local variables inside functions do not have anything to do with the variables in the the workspace. However, to save time, the variables in the argument to a MATLAB function call will not be copied into another location in the memory unless they are altered inside the m-file function. This is referred to as *call by reference*. If the argument variable is found the left of an assignment MATLAB will have to copy the variable and produce a unique instantiation for use in the m-file function. This can yield substantial time delays if the arguments are large arrays.

```
>> tic; rands1(100000); toc
elapsed_time =
            3.1300
>> tic; rands2(100000); toc
elapsed_time =
            0.3900
```

File name: primes.m

```
function p = primes(n)
if n < 2,
   p = zeros(1,0);
   return,
end
p = 1:2:n;
q = length(p);
p(1) = 2;
for k = 3:2:sqrt(n)
   if p((k+1)/2)
      p(((k*k+1)/2):k:q)=0;
   end
end
p = p(p>0);
```

File name: mysum.m

```
function y = mysum(x)
x(1) = x(1);
y = sum(x);
```

```
>> tic; sum(1:1e6); toc
elapsed_time =
            0.1100
>> tic; mysum(1:1e6); toc
elapsed_time =
            0.2200
```

When MATLAB allocates memory for a new array it is generally assumed that the array will contain real floating point numbers. Images are often quite large but commonly only consist of integer values with limited range. The example to the right illustrates how the MATLAB test image mandrill can be compressed using an unsigned integer data type. The integer data types available in MATLAB are compared to the default double precision data type in the table below

Class	Numerical range	bytes
uint8	$0 \ldots 255$	1
uint16	$0 \ldots 65535$	2
uint32	$0 \ldots 2^{32} - 1$	4
int8	$-128 \ldots 127$	1
int16	$-32768 \ldots 32767$	2
int32	$-2^{31} \ldots 2^{31} - 1$	4
double	`-realmax..realmax`	8

Consider an assignment when you are interested in determining which out of any two elements in a vector are almost equal. The first approach to such a problem would be to loop over the elements in the vector testing all possible pairs for closeness, as shown in the m-file pairtest1.m to the right. The second approach in pairtest2.m tests all possible pairings at once, utilizing the vector notation in MATLAB. In order to do this efficiently, the index vectors i and j are constructed. Here, the sparse data structure is utilized to generate these index vectors. The function find determines the location of the nonzero elements in a sparse matrix, while tril extracts the lower triangular part of a matrix, setting the upper triangular part to zero.

Both implementations yield the same result, but the first one is only feasible for vectors up to a length of a few hundred entries. The second implementation is superior to the first one in several aspects, regarding pre-allocation of memory, the elimination of the for-loops and utilization of the sparse data structure.

```
>> load mandrill
>> max(X(:))
ans =
   220
>> min(X(:))
ans =
   1
>> Xu = uint8(X);
>> all(X(:)==Xu(:))
ans =
   1
>> whos X Xu
  Name Size      Bytes  Class

  X   480x500 1920000  double array
  Xu  480x500  240000  uint8 array

Grand total is 480000 elements
using 2160000 bytes
```

File name: pairtest1.m

```
function [k,l]=pairtest1(x,dist)
k = []; l = [];
n = length(x);
for i = 1:n-1
  for j = i+1:n
    if (abs(x(i)-x(j))<dist)
      k = [k;i]; l = [l;j];
    end
  end
end
```

File name: pairtest2.m

```
function [k,l]=pairtest2(x,dist)
n = length(x);
[i,j]=find(sparse(tril(...
         ones(n,n)-speye(n))));
[k,l]=find(abs(x(i)-x(j))<dist);
```

```
>> x = rand(200,1);
>> tic;pairtest1(x,0.1);toc
elapsed_time =
   2.3000
>> tic;pairtest2(x,0.1);toc
elapsed_time =
   0.0600
```

The MATLAB function **profile** is very convenient for finding the bottleneck in personally developed m-files. The example to the right shows how the profiler works, the command **profile report** will start an html-browser containing hyper-linked information timing the different sections in the function **rands1**. The result is show in Figure 14.3.

```
>> profile on
>> rands1(100000);
>> profile report
```

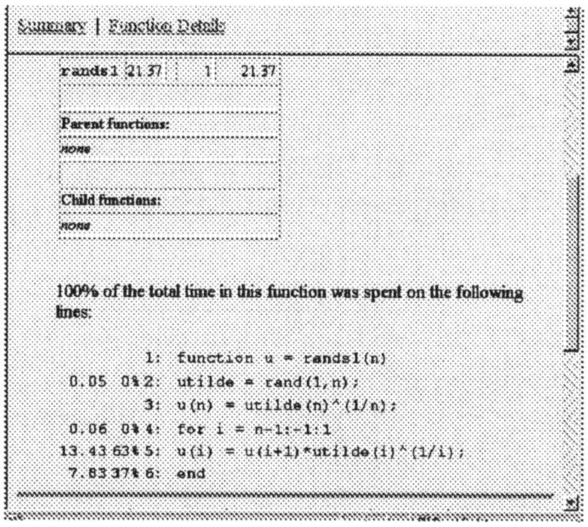

Figure 14.3 An example of a report generated by the tool **profile**.

MATLAB strives to optimize the interpreter to get faster code, and R13 introduced so called *performance acceleration*. The explicit goal is to make MATLAB as fast as C or Fortran. At this stage, R13 offers increased speed in self-contained **for** loops (not calling any m-files inside the loop, only built-in functions).

15   Calling C-routines from MATLAB

The previous section illustrated how to localize the computationally critical parts of your MATLAB code using **profile**. The section also gave several suggestions on how to improve the computational efficiency of the matlab code and decrease its execution time. However, if the computational performance is far from acceptable even with vectorized code and smart memory handling one may have to resort to implementing the algorithm in a compiled language. MATLAB provides a nice utility to call compiled Fortran or C-routines and use them just like if they were part of the MATLAB system. In Release 12 and higher of MATLAB, Java programs can also be invoked directly from the command line.

Special routines built to be executed and called from within MATLAB are called **mex**-files. These files will have different extensions depending on the platform MATLAB is running on. For instance, in Windows the mex-files have extension **.dll**, on Sun Solaris they have extension **.mexsol**. In order to build mex-files, the platform MATLAB is operating on must have a compiler installed. Most unix-platforms will have a compiler available, you will have to investigate what options you have on your system for compiling mex-files. For guidelines regarding setting up your system for mex-file compilation see the extensive on-line help for the command **mex**. This section will illustrate how to write mex-files in the C programming language, we are using a PC-platform which will invoke the Lcc-compiler that follows with the MATLAB distribution. In general, mex-file compilation is quite convenient in MATLAB and no options need to be altered. The mex-file of this section was generated using the standard option settings from the installation of MATLAB, we merely developed the C-routine and issued the command **mex** in order to generate the **.dll**-file.

Consider the **pairtest**-example from Section 14. The routine below is a C-implementation of this example developed for mex-file compilation. We have split the routine **pairtest3.c** in several sections in order to explain the details of the implementation.

File name: **pairtest3.c** part 1

```
#include "mex.h"
#include <math.h>

/* pairtest3.c - example of C-MEX file */

void mexFunction( int nlhs, mxArray *plhs[],
                  int nrhs, const mxArray *prhs[] )
{
   double *x, *l, *k, *kcpy, *lcpy;
   double d;
   int    mrows,ncols, i, j, p = 0;
```

The first section contains the inclusion of the **mex.h** header-file that defines a number of routines, for example to allocate memory for MATLAB arrays. The main function is called **mexFunction()** and always has four input arguments and no return arguments. The arguments are used to send data to the C-routine and to send back the result of the computation. The arguments are the number of output parameters in the call to the function, a pointer to the output parameters, the number of input parameters and a pointer to the input parameters, respectively. The main function head is followed by a number of declarations of local variables.

File name: `pairtest3.c` part 2

```
/* Check for proper number of arguments. */
if(nrhs!=2) {
  mexErrMsgTxt("Two inputs required.");
} else if(nlhs>3) {
  mexErrMsgTxt("Too many output arguments");
}

/* The second input must be a noncomplex scalar double.*/
mrows = mxGetM(prhs[1]);
ncols = mxGetN(prhs[1]);
if( !mxIsDouble(prhs[1]) || mxIsComplex(prhs[1]) ||
    !(mrows==1 && ncols==1) ) {
  mexErrMsgTxt("Second input must be a noncomplex scalar double.");
}

/* The first input must be a scalar double array.*/
mrows = mxGetM(prhs[0]);
ncols = mxGetN(prhs[0]);
if( !mxIsDouble(prhs[0]) || !((mrows>=1 && ncols==1) ||
    (mrows==1 && ncols>=1)) ) {
  mexErrMsgTxt("First input must be a double array.");
}
```

The second part of the routine checks the validity of the input and output arguments. Here, several special routines from the mex-library are utilized for extracting information about the MATLAB data provided by the call to the function.

File name: `pairtest3.c` part 3

```
/* Convert input array to column array */
if( mrows < ncols )
  mrows = ncols;

/* Allocate memory for maximum number of pairings */
k = (double*)malloc(sizeof(double)*mrows*(mrows-1)/2);
l = (double*)malloc(sizeof(double)*mrows*(mrows-1)/2);

/* Assign pointers to input. */
x = mxGetPr(prhs[0]);
d = mxGetScalar(prhs[1]);
```

After checking that the input parameters are correct in size and type, we determine the length of the input vector `x` and store it in the local variable `mrows`. In order to store the possible pairings we allocate memory for storing every pairing possible, which for example will happen when calling the function with a vector containing the same value in all entries. Pointers to the input parameters are also extracted using function utilities declared in `mex.h`.

File name: `pairtest3.c` part 4

```
/* The actual routine */
for (i=0; i<mrows-1; i++)
  for (j=i+1; j<mrows; j++)
    if (fabs(x[i] - x[j]) < d) {
      k[p] = i;
      l[p] = j;
      p = p + 1;
    }
```

The fourth part of the function contains the actual code, note that this code is very similar to `pairtest1.m` from Section 14. The parameter p keeps track of the actual number of pairings found by the routine.

File name: `pairtest3.c` part 5

```
/* Create two matrices for the return argument. */
plhs[0] = mxCreateDoubleMatrix(p[0],1, mxREAL);
plhs[1] = mxCreateDoubleMatrix(p[0],1, mxREAL);

/* Copy result to output matrices */
kcpy = mxGetPr(plhs[0]);
lcpy = mxGetPr(plhs[1]);
for(i = 0; i < p; i++) {
  kcpy[i] = k[i];
  lcpy[i] = l[i];
}

/* Free locally allocated memory */
free(k);  free(l);
}
```

The final part of the C-routine allocates memory for the output parameters for the p pairings. The routine copies the information from the temporary vectors k and l and frees their part of the dynamically allocated memory.

A guided tour 37 (Calling C-routines from MATLAB)

With the installation of MATLAB used to develop this section, the mex-function invokes the compiler Lcc that is bundled with the PC-version of MATLAB. The compiler generates the dynamically linked library `pairtest3.dll` which is executed when calling `pairtest3` from the command line inside MATLAB. As seen, the compiled C-routine is more than 20 times faster than the optimized m-file `pairtest2.m` developed in Section 14.

```
>> mex paritest3.c
>> x = rand(2000,1);
>> tic; [k,l]=pairtest2(x,0.1); toc
elapsed_time =
    12.2500
>> tic; [k,l]=pairtest3(x,0.1); toc
elapsed_time =
    0.5400
```

Mex-files can also be constructed as SIMULINK S-functions. Please refer to the on-line help for guidelines, examples and information on syntax for SIMULINK mex-files.

Part 3

APPLICATIONS OF MATLAB

16 Calculus

Contents:
Functions in matlab:
diff, polyder, polyint, conv, deconv, residue
Functions in @sym:

sym, syms, diff, int, taylor, symsum
simple, simplify, factor, expand

A guided tour 38 (Differentiation and integration of polynomials)

One class of functions that can easily be differentiated and integrated symbolically is polynomials. The following simple rules apply:

$$p(x) = \sum_{k=0}^{n} p_k x^k \Rightarrow$$

$$\frac{dp(x)}{dx} = \sum_{k=1}^{n} p_k k x^{k-1}$$

$$\int p(x)dx = c + \sum_{k=0}^{n} p_k \frac{1}{k+1} x^{k+1}$$

and these are implemented in **polyder** and **polyint**, respectively.

```
>> p=[1 2 3 4]
p =
     1     2     3     4
>> pd=polyder(p)
pd =
     3     4     3
>> pdi=polyint(pd,4)
pdi =
     1     2     3     4
```

We have already seen examples on how to multiply polynomials using `conv`, in guided tour 12. The inverse operation of *deconvolution* is interesting for several purposes. One is for polynomial equation solving, where we can guess one root. Consider the equation

$$x^3 - x^2 + x - 1 = 0.$$

We immediately see that $x = 1$ is one root, but the other two ones are perhaps not so easy. For that purpose, we divide the polynomial and find that

$$\frac{x^3 - x^2 + x - 1}{x - 1} = x^2 - 1,$$

which has the roots $x = \pm i$. We can verify the result by multiplying the first order polynomials and check with the original polynomial.

```
>> p=[1 -1 1 -1];
>> p1=[1 -1];
>> q=deconv(p,p1)
q =
     1     0     1
>> p2=[1 -i];
>> p3=deconv(q,p2)
p3 =
     1.0000     0 + 1.0000i
>> conv(p1,conv(p2,p3))
ans =
     1    -1     1    -1
```

Another class of functions that can be differentiated with numerically simple methods is rational polynomials. We here find that

$$\frac{d}{dx}\frac{1}{x - 0.5} = -\frac{1}{x^2 - x + 0.25}$$

```
>> b=1; a=[1 -0.5];
>> [bd,ad]=polyder(b,a)
bd =
     -1
ad =
     1.0000    -1.0000     0.2500
```

We next turn our interest to the *Symbolic toolbox* (`help symbolic`), where a lot more can be done by using symbolic computations. This toolbox uses the *Maple* kernel for all symbolic steps, but the function names are overloaded similar MATLAB functions, and the syntax is basically the same as described in Section 10 about function of functions. This means that experienced MATLAB users should not have any problems in quickly acquiring sufficient knowledge to start to solve problems similar to the ones given in basic calculus courses.

A guided tour 39 (Symbolic calculus)

Similar to functions of functions, we first need to understand how functions and symbolic objects are created, and the difference to its numerical counterparts. Consider first `pi` which in MATLAB is pre-defined with 15 decimals. Despite of this, the numerical inexactness is easy to pin-point as illustrated to the right.

If we, however, defined `pi` as symbolic, it is treated as the symbolic π and the Maple kernel basically stores all known properties about π that one can find in mathematical tables. The answer to $\sin(\pi)$ is thus *defined* to be 0, rather than computed to be 0.

```
>> sin(pi)
ans =
     1.2246e-16
>> pi=sym('pi')
pi =
pi
>> sin(pi)
ans =
0
```

We next define x to be symbolic, and we can define the function $f(x)$ similar to functions of functions, with the difference that we do not need to use `inline`.

We start with the polynomial example just considered, and note that the result is here a symbolic expression rather than a vector representing the polynomial function. Thus, the result is easier to interpret. The same holds for integration, where we also define the integration interval and get a symbolic answer:

$$\int_a^b (3x^3 + 4x^2 + 3x)dx =$$
$$b^3 - a^3 + 2b^2 - 2a^2 + 3b - 3a$$

The result becomes non-trivial when considering the function

$$f(x) = \arccos(x) \Rightarrow$$
$$\frac{df(x)}{dx} = -\frac{1}{\sqrt{1-x^2}}$$
$$\int f(x)dx = x\arccos(x) - \frac{1}{\sqrt{1-x^2}}$$

The plot function `ezplot` (pronounce it as 'easy plot') is convenient for quickly examining functions. It basically does

```
>> fplot(char(f),[-1 1])
>> title(char(f))
```

and the plot interval is chosen with at least a little intelligence. Here `char` is over-loaded in the symbolic toolbox and converts the function into a text string.

```
>> x=sym('x')
x =
x
>> f=x^3+2*x^2+3*x+4
f =
x^3+2*x^2+3*x+4
>> fd=diff(f)
fd =
3*x^2+4*x+3
>> fdi=int(fd)
fdi =
x^3+2*x^2+3*x
>> fdi=int(fd,0,1)
fdi =
6
>> syms a b
>> fdi=int(fd,a,b)
fdi =
b^3-a^3+2*b^2-2*a^2+3*b-3*a
>> f=acos(x)
f =
acos(x)
>> fd=diff(f)
fd =
-1/(1-x^2)^(1/2)
>> fi=int(f)
fi =
x*acos(x)-(1-x^2)^(1/2)
>> subplot(311), ezplot(f)
>> subplot(312), ezplot(fd)
>> subplot(313), ezplot(fi)
```

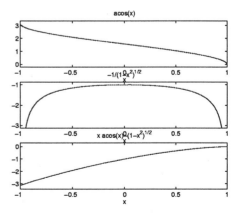

A *Taylor expansion* is computed by `taylor`, for instance the first five terms of the exponential are

$$e^x \approx 1 + x + \frac{x^2}{2} + \frac{x^3}{6} + \frac{x^4}{24} + \frac{x^5}{120}$$

for small x. A nth order expansion is returned with `taylor(f,n)`. This is indeed a *Maclaurin expansion*, and if the Taylor expansion around a point a is sought, do `syms a, taylor(f,a)`.

Also check the GUI `taylortool`.

```
>> f=exp(x);
>> disp(taylor(f))
1+x+1/2*x^2+1/6*x^3+1/24*x^4+
1/120*x^5
```

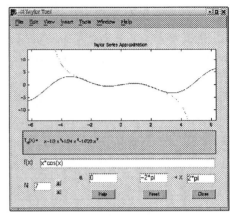

The sum of various *series* can be computed. symbolically by `symsum`. We have that

$$\sum_{k=0}^{n} k = \frac{(n+1)^2}{2} - \frac{n}{2} - \frac{1}{2}$$

$$\sum_{k=0}^{n} k^2 = \frac{(n+1)^3}{3} - \frac{(n+1)^2}{2} + \frac{n}{6} + \frac{1}{6}$$

$$\sum_{k=0}^{n} k^3 = \frac{(n+1)^4}{4} - \frac{(n+1)^3}{3} + \frac{(n+1)^2}{4}$$

The limit of a function is computed by `limit`. For instance, here we verify that

$$\lim_{x \to 0} \frac{\sin(x)}{x} = 1$$

$$\lim_{x \to \infty} \left(1 + \frac{2t}{x}\right)^{3x} = e^{6t}$$

Sooner or later the need to rewrite symbolic expressions occurs. For instance, the series above are in mathematical tables given in factorized forms, for instance

$$\sum_{k=0}^{n} k^3 = \frac{n(n+1)(2n+1)}{6}$$

```
>> syms k n
>> disp(symsum(k,0,n))
1/2*(n+1)^2-1/2*n-1/2
>> disp(symsum(k^2,0,n))
1/3*(n+1)^3-1/2*(n+1)^2+1/6*n+1/6
>> disp(symsum(k^3,0,n))
1/4*(n+1)^4-1/2*(n+1)^3+
1/4*(n+1)^2
```

```
>> syms x t; f=sin(x)/x;
>> limit(f,x,0)
ans =
1
>> f=(1+2*t/x)^(3*x);
>> limit(f,x,inf)
ans =
exp(6*t)
>> factor(symsum(k^2,0,n))
ans =
1/6*n*(n+1)*(2*n+1)
>> expand((x+3)*(x+1))
ans =
x^2+4*x+3
>> simplify(sin(x)^2 + cos(x)^2)
ans =
1
```

There are plenty of functions to simplify expressions with respect to trigonometry and algebra. Generally, one can try the high-level function **expand** on complicated expressions, and if that does not work, go on with various low-level functions. See **help symbolic** for a list of alternatives. The function **simplify** is a macro for trying different such tricks.

17 Data interpolation

> **Contents:**
> **Functions in matlab:**
> interp1, polyfit, spline,
> interp2, meshgrid, griddata, delaunay, voronoi
> **Functions in signal:**
>
> interp, decimate, resample

Interpolation is used for table lookup in many cases where either a function $y = f(x)$ or $z = f(x, y)$ is hard or impossible to compute exactly (like the sinusoid in guided tour 40 below), or the function is measured by physical sensors. In the two-dimensional case, the data (x, y) may either be given on a regular grid, or at random (as in the guided tour 41 below) or irregularly defined points. These two cases require different tools, where the latter is far more complicated.

Another application of interpolation is for sampled signals, where the grid is defined by the sampling interval. This case is covered in Section 25. The polynomial fit implemented in **polyfit** is based on the least squares principle similar to guided tour 46, and is not examined in this section.

A guided tour 40 (1D interpolation)

The function **sin** can efficiently be implemented as a table lookup. This means that we pre-compute $Y = \sin(X)$ in a number of points, for simplicity we consider the interval $X \in [0, 2\pi]$, and then interpolate the sought value(s) $y = \sin(x)$. The function **mysin** does this, where the design parameters are how dense the grid on $[0, 2\pi]$ is and how to interpolate. Note that for $x \notin [0, 2\pi]$, the remainder when dividing with 2π can be used.

There are three ways to interpolate in a function with **interp1**: nearest value, linear and cubic. Using linear interpolation, the plot cannot be distinguished from the true one.

File name: **mysin.m**

```
function y=mysin(x,arg,N);
if nargin<2
    arg='linear';
end
if nargin<3
    N=10;
end

X=linspace(0,2*pi,N);
Y=sin(X);
y=interp1(X,Y,rem(x,2*pi),arg);
```

```
>> x=(0:0.1:4*pi)';
>> y=sin(x);
>> opt={'linear','cubic',...
            'nearest'};
>> yl=mysin(x,opt{1});
>> yc=mysin(x,opt{2});
>> yn=mysin(x,opt{3});
>> plot(x,[y yl yc yn])
>> title('Interpolation using...
       N=10 points')
>> legend(opt)
```

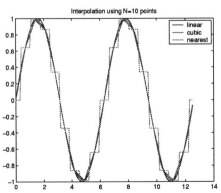

Interpolation using N=10 points

To optimize performance, we must trade off complexity to performance. Here we try to find the best combination of grid density (N points) and interpolation, by plotting computation time versus error as parametric functions of N. That is, the plot shows $(e(N), t(N))$. The three lines represent the three options for interpolations. Note that some outliers in time are always obtained due to operating system activity. To avoid this, the simulation can be repeated a number of times, and the minimum time plotted. It appears that linear interpolation with a large table is the best alternative.

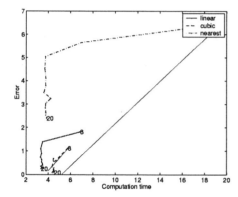

```
>> x=(0:0.01:4*pi)';
>> tic,y=sin(x);tnorm=toc;
>> for N=8:20;
   tic,yl=mysin(x,opt{1},N);
   tl(N)=toc/tnorm;
   tic,yc=mysin(x,opt{2},N);
   tc(N)=toc/tnorm;
   tic,yn=mysin(x,opt{3},N);
   tn(N)=toc/tnorm;
   errl(N)=norm(yl-y);
   errc(N)=norm(yc-y);
   errn(N)=norm(yn-y);
end
>> plot(tl(8:20),errl(8:20),'.-')
>> hold on
>> text(tl([8 20]),errl([8 20]),...
        {'8','20'})
>> plot(tc(8:20),errc(8:20),'.--')
>> text(tc([8 20]),errc([8 20]),...
        {'8','20'})
>> plot(tn(8:20),errn(8:20),'.-.')
>> text(tn([8 20]),errn([8 20]),...
        {'8','20'})
>> hold off
>> legend(opt)
>> xlabel('Computation time')
>> ylabel('Error')
```

A guided tour 41 (2D interpolation)
Interpolation in two dimensions is considerably harder than in one dimension if the data are not monotonically ordered. In one dimension, one can always do **sort** on x, and apply interpolation. This is generally not possible in two dimensions.

The example here randomizes points (x, y) from a Gaussian distribution. That is, the data is not organized in any way. The first step in interpolation, is to find all triangles defined by the points. This algorithm to perform this is called *Delaunay triangulation*. Interpolation is then performed on the triangle the point belongs to.

To compute the area the triangles support, the *convex hull* is used. It is defined as a *convex set* (the straight line between any two points in the set must then belong to the set) covering all points.

```
>> N=50; xy=randn(2,N);
>> x=xy(1,:)';
>> y=xy(2,:)';
>> t=delaunay(x,y);
>> trimesh(t,x,y)
>> hold on
>> ind=convhull(x,y);
>> h=plot(x(ind),y(ind));
>> set(h,'Linewidth',3)
```

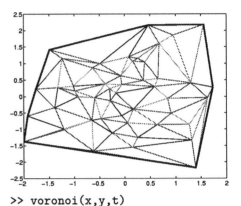

```
>> voronoi(x,y,t)
```

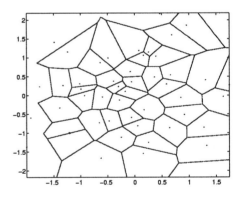

Consider the example of a telecom operator establishing cellular radio communication business in a city. He finds that the number of calls decays radially from the city center, here located at the origin, and he puts more antennas closer to the center than in the suburbs. Let the points in the vectors x,y denote the antenna positions. How do we compute the cells that each antenna cover? Assume for simplicity that only the distance influences which antenna a mobile phone is connected to. Then the cell map can be computed by just one line in MATLAB using **voronoi**.

In some applications, we can sample a two-dimensional function at random points. Here we assume that $z = f(x, y)$ is obtained from the two-dimensional Gaussian distribution. To plot the function, it has to be re-sampled on a uniform grid. The first step is thus to compute a grid using **meshgrid**, and then the function is interpolated using **griddata**, and the **mesh** function can be used. Note that the Delaunay interpolation is performed implicitly here, and that the interpolation is performed on one of the triangles, as a weighted sum of the corners' values.

To interpolate the other way around, **interp2** can be used. For instance, **zi=interp2(Xgrid,Ygrid,Z,xgrid,ygrid]** should yield an approximative recovery of **z**.

```
>> [xgrid ygrid]=meshgrid(x,y);
>> z=exp(-0.5*xgrid.^2-...
           0.5*ygrid.^2);
>> X=-3:0.2:3;
>> Y=-3:0.2:3;
>> [Xgrid Ygrid]=meshgrid(X,Y);
>> Z=griddata(x,y,z,Xgrid,Ygrid);
>> mesh(X,Y,Z)
```

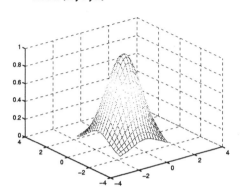

18   Linear Algebra

Contents: Two-dimensional graphical illustrations of basic concepts
Functions:
`eig, svd, norm, poly`

A standard basic course in linear algebra is organized as follows:

- It starts with geometry and introduces vectors, scalar products, subspaces and solving equation systems.

- One track develops the theory of least squares solutions for over-determined equation systems, with applications.

- Another track works with eigenvectors and terminates in the spectral theorem with applications.

A unifying tool for these latter two problems is matrix factorization, often separated from the basic course and taught in courses in numerical analysis.

Figure 18.1 illustrates the outline of this section, where the singular value decomposition (SVD) is presented in the end as one alternative to solve all numerical tasks. This section aims at illustrating the main results geometrically in $\mathcal{R}^2$.

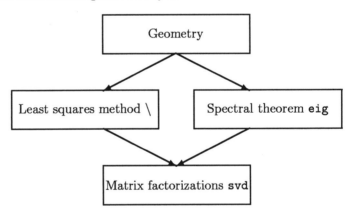

Figure 18.1   Outline of the section, where plots in $\mathcal{R}^2$ are used for illustrating the theory.

A guided tour 42 (Projections and rotations)
An ortho-normal (ON) basis for $\mathcal{R}^2$ is most naturally represented by the vectors $e_1 = (1,0)$ and $e_2 = (0,1)$. A plot of these two vectors is a bit awkward in the long run. It would require a line like `plot([0 e1(1)],[0 e1(2)])`.

```
>> e1=[1,0]'; % Base vector 1
>> e2=[0,1]'; % Base vector 1
>> e=[e1 e2]; % ON basis
```

An m-file for a vector plot in $\mathcal{R}^2$ is given to the right. It is general in that it can plot several vectors organized column-wise at the same time. The line marker is also given as an optional input argument. We can use the function to illustrate the ON basis in 2D.

File name: `vecplot.m`

```
function vecplot(e,m)
% e=Vector or vectors as
%   columns in a matrix
% m=Marker style (default '-b')
if nargin<2; m='-b'; end
for i=1:size(e,2);
    plot([0 e(1,i)],[0 e(2,i)],m)
    hold on
end
hold off
```

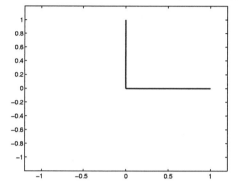

```
>> vecplot(e)
>> axis([-1.2 1.2 -1.2 1.2])
```

We also write a separate m-file for drawing lines between arbitrary points in $\mathcal{R}^2$. We have that `vecplot(e)` produces the same result as `vecvecplot(zeros(size(e)),e)`.

File name: `vecvecplot.m`

```
function vecvecplot(e1,e2,m)
% Vector to vector plot
% e1=Vector or vectors as
%    columns in a matrix
% e2=Vector or vectors as
%    columns in a matrix
% m =Marker style (default '-b')
if nargin<3; m='-b'; end
if size(e1,2)~=size(e2,2);
    error(['Equal number of ',...
        'columns in e1 and e2!'])
end

for i=1:size(e1,2);
    plot([e1(1,i) e2(1,i)],...
        [e1(2,i) e2(2,i)],m)
    hold on
end
hold off
```

An isometric mapping A is defined by Av, having $|Av| = |v|$ for all vectors v. That is, the matrix A preserves length. This is the case for all matrices satisfying $\det(A) = \pm 1$. We can interpret this mapping as a rotation with a possible mirroring. All rotations in $\mathcal{R}^2$ can be parameterized as the matrix A to the right. This mapping rotates any vector θ radians counter-clockwise.

```
>> theta=pi/6; % Rotate 30 deg
>> A=[cos(theta) -sin(theta);...
      sin(theta) cos(theta)];
>> det(A)
ans =
      1
```

We illustrate with rotating the ON-basis 30 degrees twice.

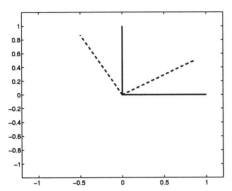

Certain plots would look much nicer with arrows instead of just lines. We can use the rotation mapping to write an arrow plot function. The arrow angle is here defined as 30 degrees, and its length is one tenth of the line.

The previous example with rotation of the ON basis is repeated with the arrow plot.

Example:

```
>> arrowplot(e,'-b')
>> hold on
>> arrowplot(A*e,'--r')
>> hold on
>> arrowplot(A^2*e,'-.g')
```

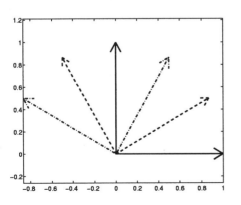

```
>> hold on
>> vecplot(A*e,'--r')
>> norm(e)==norm(A*e)
ans =
     1
>> hold on
>> vecplot(A^2*e,'-.g')
```

File name: `arrowplot.m`

```
function arrowplot(e,m)
% e=Vector or vectors as
%    columns in a matrix
% m=Marker style (default '-b')
if nargin<2; m='-b'; end
delta=0.1;  % Rel. arrow length
phi=pi/6;   % Arrow angle
% Positive rotation matrix
A1=[cos(phi) -sin(phi);...
    sin(phi) cos(phi)];
% Negative rotation matrix
A2=[cos(-phi) -sin(-phi);...
    sin(-phi) cos(-phi)];
for i=1:size(e,2);
   plot([0 e(1,i)],[0 e(2,i)],m)
   hold on
   v1=delta*A1*e(:,i);
   v2=delta*A2*e(:,i);
   plot([e(1,i) e(1,i)-v1(1)],...
        [e(2,i) e(2,i)-v1(2)],m)
   plot([e(1,i) e(1,i)-v2(1)],...
        [e(2,i) e(2,i)-v2(2)],m)
end
% Arrow looks funny without this
axis('equal')
hold off
```

A projection of the vector x on y is defined by

$$\text{proj}(x,y) = \frac{(x,y)}{(y,y)}y.$$

Here (x,y) defines the scalar product $\sum_i x^{(i)}y^{(i)}$. In MATLAB, we simply use $\texttt{x'*y}$ for computing $x^T y = (x,y)$ to utilize vectorization and avoid looping (vectors are always assumed to be column vectors here).

The projection on a plane, rather than a line, is defined analogously. Suppose the plane is the span of the *orthogonal* columns of the matrix $A = (a_1, a_2, \ldots, a_n)$. Then

$$\text{proj}(v,A) = \sum_i \frac{(v, a_i)}{(a_i, a_i)}a_i.$$

We define a function for this operation to the right. If the columns of A are not orthogonal, the orthonormalization procedure to be presented must first be applied to A.

To illustrate a projection in $\mathcal{R}^2$, two vectors are defined and the projections onto each other are computed. The vectors and projections are plotted with arrows, while the projection errors to the original vector are plotted with a line.

Note that the projection errors are always orthogonal to the line (or plane) they are projected onto:

$$(v - \text{proj}(v,A), v) = 0.$$

This is the projection theorem.

File name: `proj.m`

```
function v=proj(v1,v2);
% Computes the projection of
% the vector(s) v1 on the
% vector(s) v2
% size(v)=size(v1)
% Basic vector on vector
% projection:
% v=(v1'*v2)/(v2'*v2)*v2;
for i=1:size(v1,2)
   for j=1:size(v2,2)
     v(:,i)=(v1(:,i)'*v2(:,j))/...
        (v2(:,j)'*v2(:,j))*v2(:,j);
   end
end
```

```
>> v1=[1; 1];
>> v2=[0; 0.5];
>> arrowplot([v1 v2])
>> axis([-0.2 1.2 -0.2 1.2])
>> hold on
>> v=proj(v1,v2);
>> arrowplot(v,'--r')
>> hold on
>> vecvecplot(v,v1,':r')
>> hold on
>> v=proj(v2,v1);
>> arrowplot(v,'--g')
>> hold on
>> vecvecplot(v,v2,':g')
```

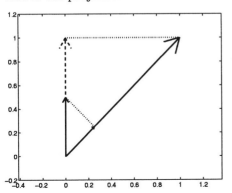

A guided tour 43 (Gram Schmidt orthonormalization)

Gram-Schmidt orthogonalization takes any linearly independent set of vectors $\{v_i\}$ and transforms it to an ON basis. It is defined by the recursion

$$\bar{e}_i = v_i - \sum_{j=1}^{i-1} \frac{(v_i, e_j)}{(e_j, e_j)} e_j,$$

$$e_i = \frac{\bar{e}_i}{|\bar{e}_i|}.$$

We write an m-file for this procedure.

If there are linearly dependent elements, this would produce a zero vector in this implementation. An alternative is to remove the vector, so the output is an ON-basis for the span of $\{v_i\}$.

File name: GS.m

```
function e=GS(v);
% Computes the Gram-Schmidt
% orthonormalization

% Initizalization
e=v(:,1)/norm(v(:,1));
for i=2:size(v,2);
  vtmp=v(:,i);
  % Orthogonalization by
  for j=1:i-1
    % repeated projection
    vtmp=vtmp-proj(v(:,i),e(:,j));
  end
  if norm(vtmp)>0
    % Normalization
    e(:,i)=vtmp/norm(vtmp);
  else
    e(:,i)=vtmp;
  end
end
```

Let's start with transforming the vectors $v_1 = (1,1)$ and $v_2 = (0, 0.5)$ to an ON basis. This is illustrated in the vector plot below.

Then an example in $\mathcal{R}^3$ is taken. The columns in the matrix A are obviously not an ON basis, since the matrix does not satisfy $A^T A = I$. This is, however, satisfied after orthonormalization.

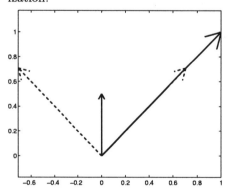

```
>> e=GS([v1 v2])
>> e =
      0.7071   -0.7071
      0.7071    0.7071
>> arrowplot([v1 v2])
>> hold on
>> arrowplot(e,'--g')
>> A=[1 1 1;1 2 3;1 2 4];
>> AON=GS(A)
AON =
      0.5774   -0.8165    0.0000
      0.5774    0.4082   -0.7071
      0.5774    0.4082    0.7071
>> det(AON)
ans =
      1.0000
>> AON'*AON
ans =
      1.0000    0.0000   -0.0000
      0.0000    1.0000   -0.0000
     -0.0000   -0.0000    1.0000
```

As another example, the linearly dependent vectors in V to the right are given to GS. The result is an ON basis, where the last vector is projected to 0.

```
V=[ [0.5;0] [1;1] [-1;2] ];
arrowplot(V)
hold on
arrowplot(GS(V),'--r')
```

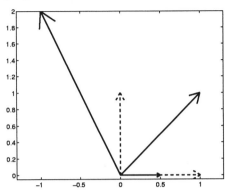

An *eigenvector* to a matrix A is defined as a vector u that satisfies

$$Au = \lambda u,$$

and λ is called the corresponding *eigenvalue*. More specifically, u is called the right eigenvector, and the left eigenvector is defined by

$$u^T A = \lambda u^T.$$

That is, a left eigenvector to A is also a right eigenvector to A^T. The most common application is for symmetric matrices, $A^T = A$, for which we can also write $u^T A = (Au)^T = \lambda u^T$. That is, the right and left eigenvectors for symmetric matrices are the same.

An interesting property of symmetric matrices is that its eigenvectors can be chosen to be orthogonal. Put all eigenvectors u in the matrix $U = [u_1, u_2, ...]$. This means that $U^T U$ is a diagonal matrix. With a proper scaling, automatically done in MATLAB, they thus constitute an ON basis, and $U^T U = I$. This is the *spectral theorem*: Any linear symmetric mapping A becomes diagonal in the basis U.

The eigenvalues are in manual calculations computed from the characteristic equation

$$\det(\lambda I - A) = 0.$$

A guided tour 44 (Eigenvalues and eigenvectors)

In the MATLAB implementation **eig**, the output is a matrix V where the columns are the *right* eigenvectors, and a diagonal matrix D, whose diagonal elements are the eigenvalues. That is, we have $AV = VD$, comprising all eigenvectors in one equation.

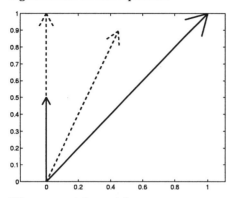

The symmetric matrix

$$A = \begin{pmatrix} 1 & 0.5 \\ 0.5 & 1 \end{pmatrix}$$

will be studied in the next couple of tours. The eigenvalue decomposition is computed, and the obtained ON basis is illustrated in a plot. Note that the order and sign of the eigenvectors are ambiguous, and may differ between different MATLAB platforms.

The characteristic polynomial is

$$\det(\lambda I - A) = (1 - \lambda)^2 - 0.25,$$

which has the solutions $\lambda = 1.5$ and $\lambda = 0.5$.

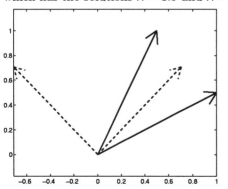

```
>> A=[v1 v2];
>> [V,D]=eig(A)
V =
            0    0.4472
       1.0000    0.8944
D =
       0.5000         0
            0    1.0000
>> A*V-V*D
ans =
       0    0
       0    0
>> arrowplot(A,'-b')
>> hold on
>> arrowplot(V,'--r')
```

```
>> A=[1 0.5;0.5 1];
>> charpol=poly(A)
charpol =
       1.0000   -2.0000    0.7500
>> lambda=roots(charpol)
lambda =
       1.5000
       0.5000
>> arrowplot(A);
>> [V,D]=eig(A)
V =
       0.7071    0.7071
      -0.7071    0.7071
D =
       0.5000         0
            0    1.5000
>> V'*V
ans =
       1.0000         0
            0    1.0000
>> hold on
>> arrowplot(V,'--r')
```

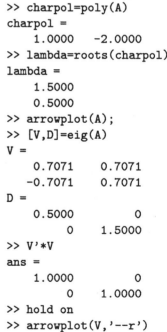

Eigenvectors provide a nice illustration of quadratic forms. A quadratic form as $x_1^2 +$

$x_1 x_2 + x_2^2 = (x_1 + 0.5 x_2)^2 + 0.75 x_2^2 > 0$ can always be written in matrix form as for this example

$$x^T A x, \quad A = \begin{pmatrix} 1 & 0.5 \\ 0.5 & 1 \end{pmatrix}.$$

Conversely, any symmetric positive definite matrix corresponds to a quadratic form. There are, however, many matrices for representing a given quadratic form. For instance,

$$x^T A x = x_1^2 + x_1 x_2 + x_2^2 = x_1^2 + (a+b) x_1 x_2 + x_2^2,$$

with $a + b = 1$, are all valid representations, corresponding to

$$A = \begin{pmatrix} 1 & a \\ b & 1 \end{pmatrix}.$$

However, if we constrain A to be symmetric ($a = b$), the representation is unique.

The eigenvalue decomposition provides a new basis in which any quadratic form becomes diagonal. We have $AV = VD$ and $A = VDV^T$, so

$$\bar{x} = V^T x \rightarrow x^T A x = x^T V D V^T x = \bar{x}^T D \bar{x} = \sum_i \lambda_i \bar{x}_i^2.$$

The interpretation of eigenvectors for symmetric matrices is thus that they define the axes of orientation of the level curve of the corresponding quadratic form. The eigenvalues further define the shape of the ellipse, since the axis $\bar{x}_i$ is proportional to $1/\sqrt{\lambda_i}$.

Another interpretation of a symmetric positive definite matrix is that it corresponds to a covariance matrix of a vector of stochastic variables. The ellipse is here interpreted as a confidence interval, and the eigenvalue decomposition provides a tool for computing confidence intervals in higher dimensions.

A guided tour 45 (Quadratic forms)

One interesting problem with *quadratic forms* is to determine the level curves, that is the values of x for which $x^T A x$ is constant. For $A = I$, the level curve is a circle. For any other positive definite matrix, we get an *ellipse* in $\mathcal{R}^2$. There are many ways to compute the shape of the ellipse. We here let x go round the unit circle, compute $x^T A x = r^2(x)$, and then plot $x/r(x)$ for all angles θ.

```
>> theta=(0:0.01:2*pi);
>> x=[cos(theta);sin(theta)];
>> r=sqrt(diag(x'*A*x));
>> hold on
>> plot(x(1,:)./r',x(2,:)./r')
>> hold off
```

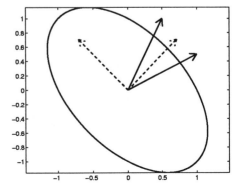

A guided tour 46 (Solving system of equations)

Consider the system of equations $Ax = b$. First, we compute the solution x in the best way numerically (we will come back to what the backslash operator implements), and then in a straightforward way. Both results are identical here.

```
>> A=[1 1;1 3];
>> b=[1.5; 3];
>> x=A\b
x =
    0.7500
    0.7500
>> x=inv(A)*b
x =
    0.7500
    0.7500
```

The vector b must lie in the span of the columns of A. A is non-singular, so there is a unique solution. We plot the columns of A, and how these are scaled by x and added to form b.

```
>> arrowplot(A,'-b')
>> hold on
>> arrowplot(b,'--r')
>> hold on
>> vecplot([x(1)*A(:,1)...
            x(2)*A(:,2)],'-.g')
>> hold on
>> vecvecplot(x(1)*A(:,1),...
   x(1)*A(:,1)+x(2)*A(:,2),'-.g')
>> hold on
>> vecvecplot(x(2)*A(:,2),...
   x(1)*A(:,1)+x(2)*A(:,2),'-.g')
```

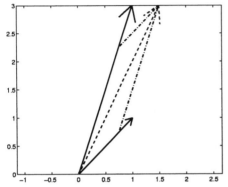

If we add one more base vector, we get an under-determined system of equations, with more unknowns than equations. In this case, there are infinitely many solutions, and the convention is to take the solution with the smallest norm x. The geometrical interpretation is that we take the cheapest linear combination to reach point b.

```
>> A=[A [-1; 1] ];
>> x=A\b
x =
        0
    1.1250
   -0.3750
>> arrowplot(A,'-b')
>> hold on
>> arrowplot(b,'--r')
>> hold on
>> vecplot([x(1)*A(:,1)],'-.g')
>> hold on
>> vecvecplot(x(1)*A(:,1),...
   x(1)*A(:,1)+x(2)*A(:,2),'-.g')
>> hold on
>> vecvecplot(x(1)*A(:,1)+...
   x(2)*A(:,2),...
   x(1)*A(:,1)+x(2)*A(:,2)+...
   x(3)*A(:,3),'-.g')
```

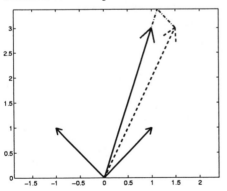

Conversely, in an over-determined system of equations there are more equations than unknowns. There is usually no exact solution here. The geometric interpretation is to take the best approximation of b in the span of A. The projection theorem here says that $\text{proj}(Ax - b, A) = 0$.

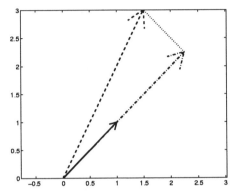

A line fit interpretation might add further insight into the examined system of equations. Suppose we have a model

$$y = \theta_1 + \theta_2 x.$$

That is, y depends linearly on x. Suppose we know two points

$$1.5 = \theta_1 + \theta_2 \cdot 1,$$
$$3 = \theta_1 + \theta_2 \cdot 3.$$

In matrix form, this is the same system of equations $Ax = b$ as above, but here the unknown vector is called θ. Geometrically, we get the line in the figure below.

```
>> A=A(:,1)
A =
     1
     1
>> x=A\b
x =
     2.2500
>> arrowplot(A,'-b')
>> hold on
>> arrowplot(b,'--r')
>> hold on
>> arrowplot([x(1)*A(:,1)],'-.g')
```

```
>> x=[1; 3];
>> y=[1.5; 3];
>> phi=[ones(size(x)) x];
>> theta=phi\y
theta =
     0.7500
     0.7500
>> plot(x,phi*theta)
>> axis([0 4 0 4])
>> hold on
>> plot(x,y,'*')
>> legend('Interpolating line',...
          'Equation constraints')
>> xlabel('x')
>> ylabel('y')
>> hold off
```

A more interesting interpretation and more practically relevant problem arises for over-determined system of equations. Here we have more points to fit the line to, and, generally, there is no perfect line interpolating all points. The least squares principle says that we should take the line that minimized the sum of squared errors. This is what is computed automatically by the backslash operator.

```
>> x=[1;   3; 2;   4;    1.4];
>> y=[1.5; 3; 2.5; 3.5; 2  ];
>> phi=[ones(size(x)) x];
>> theta=phi\y
theta =
    1.0483
    0.6367
>> phi=[1 0; 1 5];
>> plot([0 5],phi*theta);
>> hold on
>> plot(x,y,'*')
>> legend('Interpolating line',...
            'Equation constraints')
>> xlabel('x')
>> ylabel('y')
>> hold off
```

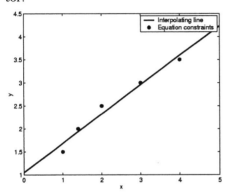

The singular value decomposition (SVD) is a numerical procedure useful for virtually any computation in linear algebra. The SVD factorizes a matrix of arbitrary dimension $n \times m$ as

$$A = USV^T \tag{18.1}$$

where U is an $n \times n$ orthonormal matrix V is an $m \times m$ orthonormal matrix and S is an $n \times m$ diagonal matrix (complemented with zero rows in the bottom or columns to the right if $n \neq m$). An orthonormal square matrix satisfies $U^T U = UU^T = I$. With a reliable algorithm, many other algorithms can be eliminated, as **inv**, **eig** and the backslash operator.

A guided tour 47 (SVD for equation solving)

We have by examples shown what the back-slash operator computes, but not how it computes it. The key in many numerical problems is to use clever factorizations of a matrix, and the singular value decomposition (SVD) is the most general factorization. We have

$$A = USV^T \Rightarrow$$
$$AVS^{-1}U^T = USV^TVS^{-1}U^T$$
$$= USS^{-1}U^T = UU^T = I.$$

Inversion of a matrix A breaks down to inversion of a diagonal matrix S, which can be done most often without numerical problems. The example to the right is well-conditioned, so all alternatives give the same result.

```
>> A=[1 1;1 3];
>> b=[1.5; 3];
>> [U,S,V]=svd(A)
U =
     0.3827   -0.9239
     0.9239    0.3827
S =
     3.4142         0
          0    0.5858
V =
     0.3827   -0.9239
     0.9239    0.3827
>> invA=V*inv(S)*U';
>> x=invA*b
x =
     0.7500
     0.7500
>> x=A\b
x =
     0.7500
     0.7500
```

A guided tour 48 (SVD for eigenvalue decomposition)

SVD can be used for eigenvalue decomposition of symmetric matrices $A = A^T$. Symmetry gives $U = V$, so

$$A = USU^T \Rightarrow AU = USU^TU = US,$$

which is the definition of an eigenvalue decomposition.

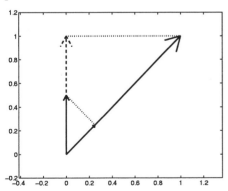

```
>> A=[1 0.5;0.5 1];
>> [Evec,Eval]=eig(A)
Evec =
     0.7071    0.7071
    -0.7071    0.7071
Eval =
     0.5000    1.5000
>> [U,S,V]=svd(A)
U =
     0.7071   -0.7071
     0.7071    0.7071
S =
     1.5000         0
          0    0.5000
V =
     0.7071   -0.7071
     0.7071    0.7071
```

A guided tour 49 (SVD for least squares problems)

We will here construct an ill-conditioned least squares problem, illustrating that factorization beats the direct solution. The least squares solution to $Ax = b$ can be written $x = (A^T A)^{-1} A^T b$. Already by forming the so called normal equations $A^T A x = A^T b$, we get numerical problems here. The factorization approach works fine. The matrix S is here a thin diagonal matrix, which means that the lower part is all filled with zeros. What is the exact solution to this problem?

```
>> delta=1e-10;
>> A=[1 2;2 4;3 6+delta];
>> b=[1;1;1];
>> [U,S,V]=svd(A)
U =
      0.2673   -0.3586   -0.8944
      0.5345   -0.7171    0.4472
      0.8018    0.5976    0.0000
S =
      8.3666        0
           0   0.0000
           0        0
V =
      0.4472   -0.8944
      0.8944    0.4472
>> x=inv(A'*A)*A'*b
Warning: Matrix is singular to
working precision.
x =
    Inf
    Inf
>> bbar=U'*b;
>> x=V*inv(S(1:2,1:2))*bbar(1:2,:)
x =
    1.0e+10 *
      1.6000
     -0.8000
>> x=A\b
x =
    1.0e+10 *
      1.6000
     -0.8000
```

Exercise 57

Solve the system of equations

$$
\begin{aligned}
x_1 & +x_2 & & = 2 \\
& x_2 & +x_3 & = 3 \\
x_1 & & +x_3 & = 4
\end{aligned}
$$

Exercise 58

Solve the system of equations

$$
\begin{aligned}
x_1 & -x_2 & +x_3 & -x_4 & = 1 \\
& x_2 & +x_3 & +x_4 & = 2 \\
x_1 & x_2 & +3x_3 & +x_4 & = 5
\end{aligned}
$$

Exercise 59

Consider the equation

$$x_1 + ax_2 \;=\; b.$$

If $x = (x_1, x_2)^T$ were known, we could illustrate the equation in an (a, b) plot as a straight line. Suppose now that two points on this line are given,

$$x_1 + 1x_2 \;=\; 2$$
$$x_1 + 2x_2 \;=\; 3.$$

That is, we have two equations for two unknowns. Solve the equation system, plot the straight line and the given points in the same plot. Suppose we get another point

$$x_1 + 4x_2 \;=\; 6$$

Solve the over-determined equation system and illustrate the best straight line fit in a plot.

Exercise 60

Let $u = (1, 2, -1)$ and $v = (3, 1, 1)$. Compute $|u+v|$, the projection of u on v and the angle between u and v (remember the formula $(u, v) = |u| \cdot |v| \cdot \cos(\phi)$.

Exercise 61

Let e_1, e_2 be an ON basis in $\mathcal{R}^2$. Determine the transformation matrix T such that $e_i^T = Te_i$ gives a new ON basis, which is the original one rotated 45 degrees.

Suppose the point $x = (2, 1)$ is given in the ON basis $e_1 = (1, 0)$ and $e_2 = (0, 1)$. What are the coordinates in the new basis?

Exercise 62

Determine all vectors perpendicular to $(1, 0, 1)$ and $(3, 2, -1)$.

Exercise 63

Determine the equation of the straight line on page 109 passing the points $p_1 = (1, 1.5)$ and $p_2 = (3, 3)$ by using the expression $p + tv$, where p is a point, v a vector and t a real number as a free parameter. Take for instance $p = p_1$ and $v = p_2 - p_1$.

Exercise 64

Show that the vectors y that make the system of equations

$$
\begin{array}{rrrl}
x_1 & -2x_2 & +3x_3 & = y_1 \\
3x_1 & +5x_2 & -4x_3 & = y_2 \\
5x_1 & x_2 & +2x_3 & = y_3
\end{array}
$$

solvable belongs to a plane. Hint: compute the null space and value spaces by an eigenvalue decomposition.

Exercise 65

Determine the eigenvectors of

$$A = \begin{pmatrix} 2 & 1 & 1 \\ 1 & 2 & 1 \\ 1 & 1 & 2 \end{pmatrix}.$$

Then find a new ON basis in which this linear mapping becomes diagonal. That is, find T such that TAT^{-1} is diagonal.

Exercise 66

Find a square matrix A such that

$$A^2 = \begin{pmatrix} 6 & 3 \\ 3 & 7 \end{pmatrix},$$

without using sqrtm. Hint: use SVD.

Exercise 67

Sketch the ellipse

$$x_1^2 + 4x_1x_2 + x_2^2 = 0,$$

by using the symmetric matrix A such that the equation becomes $x^T A x = 0$. Hint: determine the ellipse's axes and lengths with eig.

Exercise 68

Sketch the hyperbola

$$x_1^2 + 4x_1x_2 + x_2^2 + 2x_1 - x_2 = 0,$$

by using the symmetric matrix A such that the equation becomes $x^T A x + 2x_1 - x_2 = 0$. Hint: use eig to compute a new ON basis.

Exercise 69

Let

$$A = \begin{pmatrix} 1 & 2 & 3 \\ 2 & 1 & 2 \\ 3 & 2 & 1 \end{pmatrix}.$$

Determine the largest value of $|Ax|$ when $|x| = 1$. Hint: use eig.

Exercise 70

Use SVD to compute the least squares solution to the example on page 109.

Exercise 71

Rewrite the m-file vecplot for illustrating vectors in $\mathcal{R}^3$, so that when its input argument is a vector in $\mathcal{R}^3$, then the built-in function plot3 is used instead of plot.

19   Optimization

> **Contents:** Illustration of basic functions in the Optimization Toolbox
> **Functions:**
> fmincon, fminunc, fseminf, lsqcurvefit, lsqnonlin, linprog, quadprog

Constrained optimization

A guided tour 50 (Linear programming)

We will in this section frequently study linear inequalities on the form

$$Ax \leq b$$

Numerically, we will use

$$\begin{pmatrix} 1 & 2 \\ -1 & -1 \\ -3 & 2 \\ 2 & -4 \end{pmatrix} \begin{pmatrix} x_1 \\ x_2 \end{pmatrix} \leq \begin{pmatrix} b_1 \\ b_2 \\ b_3 \\ b_4 \end{pmatrix}. \qquad (19.2)$$

To illustrate geometrically what is going on, the unknown x to be determined is two-dimensional, and our first task is to compute the lines that each row of this system of inequalities corresponds to. If we solve each row i for x_2, we get

$$x_2 \leq \frac{b(i) - A(i,1)x_1}{A(i,2)}.$$

This works if there are no vertical lines. That is, the second column of A must be non-zero to avoid division with zero. The function to the right computes this value for a grid on x_1 between -3 to 3.

File name: conslines.m

```
function x2=conslines(A,b,x1);
% Computes constraints lines
for i=1:size(x1,1);
  x2(i,:)=[...
    (b(i)-A(i,1)*x1(i,:))/A(i,2)];
end
```

```
>> hold on
>> A=[1 2; -1 -1; -3 2; 2 -4];
>> b=[1; 2; 3; 4];
>> n=length(b);
>> x1=ones(n,1)*[-3:3];
>> x2=conslines(A,b,x1);
```

The next task is to illustrate the region $Ax \leq b$ by plotting the lines $Ax = b$. The function **consplot** loops over the row indices and plots each line with a text describing the equality.

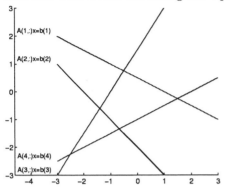

Since we still don't know where the region $Ax \leq b$ is, we add gradient arrows using **quiver** to each line. The gradient for line i is given by $A(i,:)$, which is the direction where $A(i,:)x$ increases the most.

We see that the region $Ax \leq b$ is closed in this case, and that the origin is contained in the region.

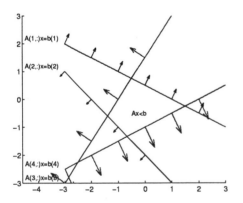

File name: `consplot.m`

```
function consplot(x1,x2)
% Plots constraint lines
for i=1:size(x1,1)
    plot(x1(i,:),x2(i,:),'-')
    text(x1(i,1)-1.5,...
        0.2+x2(i,1),...
        ['A(',num2str(i),...
        ',:)x=b(',num2str(i),')'])
end
```

```
>> hold on
>> consplot(x1,x2);
>> axis([-4.6 3 -3 3])
```

File name: `consquiver.m`

```
function consquiver(A,x1,x2)
for i=1:size(x1,1)
    for j=1:size(x1,2);
        quiver(x1(i,j),x2(i,j),...
            0.2*A(i,1),0.2*A(i,2))
    end
end
```

```
>> consquiver(A,x1,x2);
>> text(-0.5,-0.5,'Ax<b')
```

The *linear programming* problem is to maximize a linear function

$$\min_{x} f^T x$$

where f is a column vector, subject to the constraints

$$Ax \leq b.$$

We illustrate the linear function by plotting the line $f^T x = 0$, with gradient arrows in the direction f, where $f^T x$ increases.

It should be clear now where the minimum is, and one can also realize that the minimum is always in a corner of the set $Ax \leq b$ (or on a line), whenever the set is closed.

The `linprog` function in the optimization toolbox computes the solution with an iterative method. The code to the right first computes the solution, and then starts a loop to find the approximation after each iteration. First, an interior point $x^{(0)}$ has to be found. After this, we quickly move along the negative gradient to the border of the region $Ax \leq b$, and the solution is found in four iterations.

Note that the optional parameters are stored in a `struct`, initialized with default values in the function `optimset`.

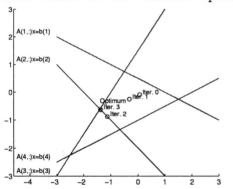

File name: `gradientplot.m`

```
function gradientplot(f,x1)
plot(x1,-x1*f(1)/f(2),'--r')
for i=1:size(x1,2);
  quiver(x1(i),...
      -x1(i)*f(1)/f(2),...
      0.5*f(1),...
      0.5*f(2),'--r')
end
```

```
>> hold on
>> consquiver(A,x1,x2);
>> consplot(x1,x2);
>> f=[2;1];
>> gradientplot(f,-2:2);
>> axis([-4.6 3 -3 3])

>> hold on
>> consplot(x1,x2);
>> options=optimset(...
      'Diagnostics','off',...
      'Display','off');
>> x=linprog(f,A,b,...
      [],[],[],[],[],options)
>> x =
      -1.4000
      -0.6000
>> plot(x(1),x(2),'*')
>> text(x(1),x(2)+0.3,'Optimum')
>> for i=0:3;
   options=optimset(...
      'maxiter',i,...
      'Diagnostics','off',...
      'Display','off');
   x=linprog(f,A,b,...
      [],[],[],[],[],options);
   plot(x(1),x(2),'o')
   text(0.1+x(1),0.1+x(2),...
      ['Iter. ',num2str(i)])
end
>> axis([-4.6 3 -3 3])
```

A guided tour 51 (Quadratic programming)

The *quadratic programming* problem is to minimize a quadratic function

$$\min_x 0.5 x^T H x + f^T x,$$

where f is a column vector and H a square matrix, subject to the constraints

$$Ax \le b.$$

Here we study

$$\min_x x^T \begin{pmatrix} 1 & 0.25 \\ 0.25 & 1 \end{pmatrix} x + \begin{pmatrix} 2 \\ 1 \end{pmatrix}^T x$$

$$= \min_x x_1^2 + 0.5 x_1 x_2 + x_2^2 + 2 x_1 + x_2$$

$$\begin{pmatrix} 1 & 2 \\ -1 & -1 \\ -3 & 2 \\ 2 & 4 \end{pmatrix} \begin{pmatrix} x_1 \\ x_2 \end{pmatrix} \le \begin{pmatrix} b_1 \\ b_2 \\ b_3 \\ b_4 \end{pmatrix}.$$

We first want to illustrate the quadratic function by plotting the level curves of $x^T H x + f^T x = c$. How to plot an ellipse to a *quadratic form* was described in guided tour 45.

Contrary to the linear case, there is no need to plot the gradient here, since a quadratic function always increases with the distance from its center point.

The center point can be found by completing the squares as

$$c = x^T H x + f^T x$$

$$(x+0.5H^{-1}f)^T H (x+0.5H^{-1}f) - 0.25 f^T H^{-1} f.$$

That is, we can compute the ellipse $\bar{x}^T H \bar{x} = c + 0.25 f^T H^{-1} f = \bar{c}$, and then move it to the center point $-0.5 H^{-1} f$.

We plot the constraint lines and two level curves in the same plot.

The function **quadprog** in the optimization toolbox iteratively computes the solution.

File name: `quadplot.m`

```
function quadplot(H,f,c)
% Plots the level curves
% to 0.5x'Hx+f'x=c
H=0.5*H;
theta=(0:0.01:2*pi);
x=[cos(theta);sin(theta)];
r=sqrt(2*diag(x'*H*x));
m=-0.5*inv(H)*f;
cbar=sqrt(c+0.25*f'*inv(H)*f);
plot(m(1)+x(1,:)./r'*cbar,...
    m(2)+x(2,:)./r'*cbar,'--r')
```

```
>> hold on
>> H=[1 0.25; 0.25 1];
>> f=[2;1];
>> consplot(x1,x2);
>> quadplot(H,f,1);
>> quadplot(H,f,-1);
>> axis([-4.6 3 -3 3])

>> options=optimset(...
      'Diagnostics','off',...
      'Display','off');
>> x=quadprog(H,f,A,b,...
      [],[],[],[],[],options)
>> x =
      -1.4000
      -0.6000
>> plot(x(1),x(2),'*')
>> text(x(1),x(2)+0.3,'Optimum')
>> axis([-4.6 3 -3 3])
```

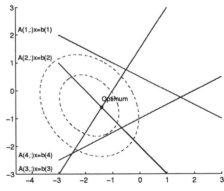

Least squares

The *linear least squares* problem appeared in guided tour 49 in Section 18. It is here revisited in a somewhat different view, with the idea to extend it to *systems of non-linear equations* and *curve fitting*.

A guided tour 52 (Linear least-squares)

Consider two of the lines in the previously studied relation $Ax = b$ in (19.2), corresponding to rows 2 and 3, respectively. These lines intersect within the plot window.

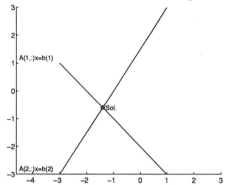

```
>> hold on
>> A=[-1 -1; -3 2];
>> b=[2; 3];
>> n=length(b);
>> x1=ones(n,1)*[-3:3];
>> x2=conslines(A,b,x1);
>> consplot(x1,x2);
>> x=inv(A)*b
x =
      -1.4000
      -0.6000
>> plot(x(1),x(2),'o')
>> text(x(1)+0.1,x(2),'Sol.')
>> axis([-4.6 3 -3 3])
```

Now, the least squares solution to the over-determined system of equations $Ax = b$ in (19.2) is computed.

Here we used the definition of *pseudo-inverse* $A^\dagger = (A^T A)^{-1} A^T$, but as stated earlier, the *backslash* operator \ is often to be preferred.

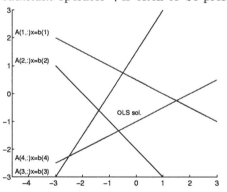

```
>> A=[1 2; -1 -1; -3 2; 2 -4];
>> b=[1; 2; 3; 4];
>> n=length(b);
>> x1=ones(n,1)*[-3:3];
>> x2=conslines(A,b,x1);
>> consplot(x1,x2);
>> x=inv(A'*A)*A'*b
x =
      -0.6299
      -0.6772
>> LSfit=(A*x-b)'*(A*x-b)
LSfit =
      21.9685
>> plot(x(1),x(2),'o')
>> text(x(1)+0.1,x(2),'LS sol.')
>> axis([-4.6 3 -3 3])
```

A guided tour 53 (Non-linear least squares)

The optimization problem becomes more involved when non-linear functions are studied. Here we want to find the x giving the solution to the equations

$$\begin{pmatrix} -1 & -1 \end{pmatrix} x = 2$$

$$x^T \begin{pmatrix} 1 & 0.25 \\ 0.25 & 1 \end{pmatrix} x + \begin{pmatrix} 2 & 1 \end{pmatrix} x =$$

$$x_1^2 + 0.5x_1x_2 + x_2^2 + 2x_1 + x_2 = 1.$$

As usual, we can convince ourselves of the plausibility of the solution by plotting the equations. Generally, a system of non-linear equations can have none, one or many solutions. In this case, we see that there are two solution, of which we find one. By testing other starting points than the origin, we will find the other one as well.

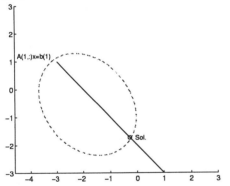

```
>> hold on
>> H=[1 0.25; 0.25 1];
>> f=[2;1];
>> c=1;
>> x1=ones(1,1)*[-3:3];
>> x2=conslines(A(2,:),b(2),x1);
>> consplot(x1,x2);
>> quadplot(H,f,c);
>> fun=inline('[ [-1 -1]*x-2;...
      x''*[1 0.25;0.25 1]*x+...
      [2;1]''*x-1]','x')

fun =
  Inline function:
  fun(x) = [ [-1 -1]*x-2;...
  x'*[1 0.25;0.25 1]*x+[2;1]'*x-1]
>> xhat=lsqnonlin(fun,[0;0])
>> xhat =
     -0.2792
     -1.7208
>> fun(xhat)
ans =
   1.0e-06 *
     0.0000
     0.3426
>> plot(xhat(1),xhat(2),'o')
>> text(xhat(1)+0.2,...
         xhat(2),'Sol.')
>> axis([-4.6 3 -3 3])
```

For over-determined systems of non-linear functions, we face an interesting optimization problem. The least squares solution is the one that minimizes the squared distance to all curves. The example here is

$$\begin{pmatrix} -1 & -1 \\ -3 & 2 \end{pmatrix} \begin{pmatrix} x_1 \\ x_2 \end{pmatrix} = \begin{pmatrix} 2 \\ 3 \end{pmatrix}$$

$$x^T \begin{pmatrix} 1 & 0.25 \\ 0.25 & 1 \end{pmatrix} x = x_1^2 + 0.5x_1x_2 + x_2^2 = 1.$$

The intersection of the two linear equations is far away from the ellipse, so the compromise is closer to the origin of the ellipse.

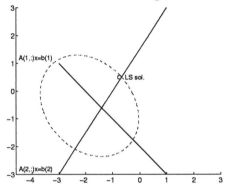

```
>> hold on
>> x1=ones(2,1)*[-3:3];
>> x2=conslines(A(2:3,:),...
                b(2:3),x1);
>> consplot(x1,x2);
>> quadplot(H,f,c);
>> fun=inline('[ [-1 -1]*x-2;...
         [-3 2]*x-3;...
         x''*[1 0.25;0.25 1]*x+...
         [2;1]''*x-1]','x')
fun =
     Inline function:
     fun(x) = [ [-1 -1]*x-2;...
         [-3 2]*x-3;...
         x'*[1 0.25;0.25 1]*x+...
         [2;1]'*x-1]
>> xhat=lsqnonlin(fun,[0;0])
xhat =
     -0.7446
      0.5021
>> fun(xhat)
ans =
     -1.7575
      0.2380
     -1.3675
>> plot(xhat(1),xhat(2),'o')
>> text(xhat(1)+0.2,...
         xhat(2),'LS sol.')
>> axis([-4.6 3 -3 3])
```

A guided tour 54 (Least squares curve fitting)

A related problem to non-linear least squares is the *curve fitting* problem. Given a parametric function

$$y = f(x; \theta),$$

find the best value of the unknown parameter θ (possibly a vector) that interpolates given data as well as possible.

We are here interested in finding a formula for the density of primes. Let p_k be prime number k. For example, $p_1 = 1$, $p_4 = 5$, $p_5 = 7$. The ratio p_k/k is the average distance between two prime numbers for all integers up to p_k. We see from the plot that around 1000, there are one prime number per ten integers approximately. How does this number evolve?

```
>> hold on
>> p=primes(1e3)';
>> ind=(1:length(p))';
>> plot(p./ind)
>> xlabel('k')
>> ylabel('p(k)/k')
>> title('Density of primes')
```

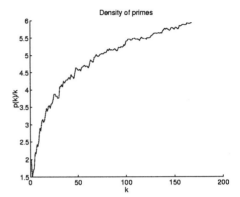

We might postulate a logarithmic growth

$$\frac{p_k}{k} = f(k, \theta) = \theta \log(k),$$

and the problem is to compute the value of θ giving the best fit. The function `lsqcurvefit` from the optimization toolbox is used, and we see that the model

$$\frac{p_k}{k} \approx 1.16 \log(k)$$

comes quite close to the data.

```
>> f=inline('th*log(k)','th','k');
>> th0=1;
>> thhat=lsqcurvefit(f,th0,...
                        ind,p./ind)
thhat =
    1.1613
>> hold on
>> plot(feval(f,thhat,ind),'--')
```

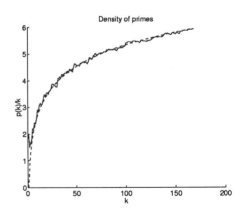

20   Numerical Accuracy and Number Representation

Contents: IEEE standard for real numbers
Functions:
eps, Inf, NaN, realmin, realmax, format long, format hex

Internally, MATLAB handles real, floating point, numbers according to the IEEE–standard for floating point arithmetic using extended double precision. Some background knowledge regarding finite precision floating point systems in general, and this standard in particular, is vital in order to interpret the results and quantify the accuracy from computations performed in MATLAB.

In the computer, any floating point number x is decomposed as $x = (-1)^s b 2^e$ and stored as the three integer values e, b, and s. The *exponent* e ranges between a negative and a positive integer, $L \leq e \leq U$, b is a positive integer stored in a t-bit *mantissa* and the sign of x is indicated by a *sign bit*, s. The precision and range of numbers x that can be stored in the computer is dependent on the number of bits used to represent b and e. The precision is dependent on the type of machine MATLAB is running on. For example, on a standard PC MATLAB makes use of eight bytes for each floating point number. Schematically, the 64 bits are divided between e and b according to:

Five laws govern the way MATLAB represents the floating point number:

- $e = U$ and $b \neq 0$ yields that $x = $ NaN (Not-a-Number)

- $e = U$ and $b = 0$ yields that $x = (-1)^s \infty$

- $L < e < U$ yields that $x = (-1)^s [1.b_1 b_2 \ldots b_t]_2 \, 2^e$

- $e = 0$ and $b \neq 0$ yields $x = (-1)^s [0.b_1 b_2 \ldots b_t]_2 \, 2^L$

- $e = 0$ and $b = 0$ yields $x = (-1)^s 0$

where $[b_0.b_1 b_2 \ldots b_t]_2 = \sum_{i=0}^{t} b_i 2^{-i}$. Hence, there are two extra symbols for representing special results from operations. The symbol Inf indicates a result outside the interval that can be stored in the computer, while NaN represents a mathematically undefined quantity. Both these symbols are predefined as constants in MATLAB.

Exercise 72
With format hex *MATLAB displays floating point numbers as they are stored in memory using hexadecimal numbers. Let $a = 1 + 2^{-n}$. Determine the smallest integer n such that $a = 1$ and the largest integer n such that $a > 1$. Verify that the mantissa length $t = 52$.*

Exercise 73
Determine the exponent range $[L, U]$. Determine U by using format hex *and study how $a = 2^n$ is stored when successfully increasing the integer n, what is the largest n such that $a < \infty$? Determine L by studying $a = (1 + 2^{-t}) 2^{-n}$ for increasing n, the largest n such that a is stored with maximal accuracy will determine L.*

Exercise 74
What is the maximal and the minimal real number that can be stored in the computer memory with maximal accuracy? Determine analytical expressions for these quantities in terms of t, L and U. Compare your results with the constants `realmin` *and* `realmax`.

Exercise 75
Numerically evaluate the following four expressions in MATLAB

$$\sin(\pi) \qquad 0 - e^{987} \qquad e^{986} - e^{987} \qquad 1 + 10^{-17}$$

and try to explain the results of the calculations.

Whenever the real number x falls between the upper and lower limits given by the MATLAB constants `realmin` and `realmax`, we have $x = \pm[1.b_1b_2\ldots b_t]_2 2^e$. The spacing between two consecutive floating point numbers that can be stored in the computer with maximal accuracy is therefore 2^{e-t}. Hence,

$$|x_{\text{MATLAB}} - x| \leq \frac{1}{2}2^{e-t}$$

and since $2^{e-1} < |x|$ it follows that the relative accuracy is given by

$$\mu = \frac{|x_{\text{MATLAB}} - x|}{|x|} \leq 2^{-t}.$$

This fact was empirically investigated in Exercise 72. On a standard PC implementation $\mu = 2^{-52} \approx 2.2 \cdot 10^{-16}$ and is the reason why `format long` shows 15 decimals.

A guided tour 55 (Floating point accuracy)
The machine constant μ is available in MATLAB as the predefined constant `eps`. For example, in MATLAB, the numerical accuracy yields that

$$1 + \mu^2 = 1$$

even though both the numbers 1 and μ^2 can be represented in the MATLAB floating point system.

```
>> format long
>> eps
ans =
     2.220446049250313e-016
>> 1+eps^2
ans =
     1
>> eps^2
ans =
     4.930380657631324e-032
```

Exercise 76
Predict the result of the computation `realmax + 10000`. *Verify your prediction using* MATLAB.

A guided tour 56 (Numerical problems in matrix algebra)
Let us return to guided tour 13 on page 28. Suppose we change the $(3,3)$-element in the matrix A to 9.

$$A = \begin{pmatrix} 1 & 2 & 3 \\ 4 & 5 & 6 \\ 7 & 8 & 9 \end{pmatrix}$$

```
>> format short
>> A(3,3)=9
A =
     1     2     3
     4     5     6
     7     8     9
```

Now, A becomes singular. This is confirmed by a zero determinant.

```
>> det(A)
ans =
      0
```

The rank of A is 2. That is, there are two linearly independent rows and columns, respectively.

```
>> rank(A)
ans =
      2
```

The linear dependency between the rows is found by elementary row operations performed by the command **rref**, reduced row echelon form. The last entry x_3 is now free, so parameterizing $x_3 = t$ and solving for the remaining entries yields that

```
>> rref(A)
ans =
      1      0     -1
      0      1      2
      0      0      0
```

$$\begin{pmatrix} 1 & 0 & -1 \\ 0 & 1 & 2 \\ 0 & 0 & 0 \end{pmatrix} \begin{pmatrix} x_1 \\ x_2 \\ x_3 \end{pmatrix} = 0 \quad \Rightarrow \quad x = t \begin{pmatrix} 1 \\ -2 \\ 1 \end{pmatrix}$$

which defines the nullspace of A since the equation $Ax = 0$ for all x in this linear space.

This can also be verified by computing the eigenvalues and eigenvectors. The eigenvector corresponding to the zero eigenvalue defines the nullspace of A. Note though that the eigenvectors computed by **eig** are scaled to have unit length. See also **null**.

```
>> [V,D]=eig(A)
V =
      0.2320      0.7858      0.4082
      0.5253      0.0868     -0.8165
      0.8187     -0.6123      0.4082
D =
     16.1168           0           0
           0     -1.1168           0
           0           0     -0.0000
```

The condition number is very large, indicating that numerical problems can occur when trying to invert A. Analytically, the condition number is infinite but the eigenvalues cannot be computed exactly. Note that the actual number displayed in such warnings may vary between different hardware platforms.

```
>> cond(A)
ans =
      8.5796e+16
```

The reciprocal condition number is therefore a better indicator of matrix singularity. It is close to zero for badly conditioned matrices and close to one for well conditioned matrices. Remember that the relative accuracy is given by the constant **eps** showing that A indeed is badly conditioned.

```
>> rcond(A)
ans =
      2.0560e-018
>> eps
ans =
      2.2204e-016
```

Despite all these indicators of singularity trying to solve the equation

$$\begin{pmatrix} 1 & 2 & 3 \\ 4 & 5 & 6 \\ 7 & 8 & 9 \end{pmatrix} x = \begin{pmatrix} 1 \\ 1 \\ 1 \end{pmatrix}$$

using matrix inversions, yields a warning message and an incorrect solution. The reciprocal condition number is shown in the warning message.

By discarding one of the equations we obtain an under-determined system of two equations and three unknowns. The remaining matrix is of full rank and the under-determined system it defines has therefore infinitely many solutions. The backslash operator can be used to determine the unique solution having minimum norm, the so-called least squares solution. The computed solution satisfies the total set of three equations, having a residual error of size close to eps.

Adding the nullspace to the least squares solution yields a parameterization of the total solution to the singular equation system

$$x = \begin{pmatrix} -0.5 \\ 0 \\ 0.5 \end{pmatrix} + t \begin{pmatrix} 1 \\ -2 \\ 1 \end{pmatrix}$$

```
>> b=[1;1;1];
>> x=inv(A)*b
Warning:
    Matrix is close to
    singular or badly scaled.
    Results may be inaccurate.
    RCOND = 2.055969e-18
x =
   -2.5000
    5.0000
   -1.5000

>> norm(A*x-b)
ans =
    9.6437
>> rank(A(1:2,:))
ans =
    2
>> x=A(1:2,:)\b(1:2,1)
x =
   -0.5000
        0
    0.5000
>> norm(A*x-b)
ans =
   1.8841e-015
>> norm(A*(x+1.24*[1;-2;1])-b)
ans =
   1.9860e-015
```

21   Statistics

> **Contents:** Illustration of random numbers and their distribution
> **Functions:**
> rand, randn, mean, std, med, hist, bar, erf

The density function for the *Gaussian distribution*, or with another name the *normal distribution*, is defined as

$$f(x) = \frac{1}{\sqrt{2\pi}\sigma}e^{-\frac{(x-\mu)^2}{2\sigma^2}},$$

where μ denotes the mean value and σ is the standard deviation. What we want to compute for a normalized ($\mu = 0$, $\sigma = 1$) Gaussian variable is

$$\text{Prob}(X < x) \leftrightarrow \int_{-\infty}^{x} \frac{1}{\sqrt{2\pi}}e^{-x^2/2}dx \qquad (21.3a)$$

$$\text{Prob}(X > x) \leftrightarrow \int_{x}^{\infty} \frac{1}{\sqrt{2\pi}}e^{-x^2/2}dx \qquad (21.3b)$$

which is referred to as the *Gaussian error function*. In the general case, we re-scale the stochastic variable as

$$\text{Prob}(\frac{X - \mu}{\sigma} < x)$$

and use (21.3).

A guided tour 57 (Statistics)

Let's start with generating a vector of random numbers from the *normal distribution*, using the **randn** function with the arguments 1000 rows and one column.

```
>> X=randn(1000,1);
```

The mean value and standard deviation should be 0 and 1, respectively. In this realization with 1000 samples the sample mean and sample standard deviation turned out to be 0.03 and 1.01, respectively. The median (if the elements in a vector is sorted, the median is the mid value) is around zero as well.

```
>> mean(X)
ans =
    0.0314
>> std(X)
ans =
    1.0096
>> median(X)
ans =
    0.0462
>> hist(X);
```

Plotting a histogram with **hist** over the samples shows us that the random numbers are centered around zero with very few samples outside the interval $[-3, 3]$.

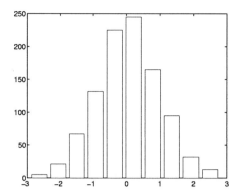

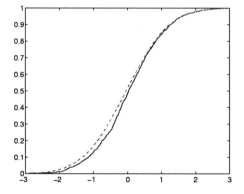

The histogram should resemble the density function in the long run. We can plot the density function and histogram in the same diagram, but we must re-scale the histogram so its integral is one (that is always true for a density function). The **bar** function can be used to plot a histogram with arbitrary scaling. Since there are 1000 x-elements and the grid width is 0.1, we scale **fhat** by $(1000 \cdot 0.1)^{-1} = 0.01$ such that its integral is 1. As we can see, the histogram and density function come quite close to each other.

To compute the probability that a random sample from the normal distribution is less than a, denoted $P(X < a)$, we can use our samples and simply count the number of outcomes that are less than a and divide by the number of samples, in this case 1000.

The probability distribution function is defined as

$$F(a) = \int_{-\infty}^{a} f(x)dx$$

and denotes the probability that a random number is less than a. We can use the **cumsum** function to integrate numerically, where the grid width 0.1 corresponds to dx in the integral. The true and estimated distribution functions are almost identical.

```
>> x=-3:0.1:3;
>> f=1/sqrt(2*pi)*exp(-x.^2/2);
>> [fhat,dum]=hist(X,x);
>> plot(x,f)
>> hold on
>> bar(x,0.01*fhat,'w')
>> hold off

>> ind=find(X<1);
>> length(ind)/1000
ans =
    0.8260

>> [Xsort,ind]=sort(X);
>> Fhat=cumsum(Xsort);
>> F=0.1*cumsum(f);
>> plot(Xsort,(1:1000)/1000,...
        'b-',x,F,'m--');
```

Now when we have computed the "true" distribution function, at least up to the resolution in the x-vector, we can use it to compute $P(X < 1)$.

There are two Matlab functions, `erf` and `erfc`, that integrate the normal density function numerically. It is however scaled in a tricky and non-obvious way:

$$\text{erf}(\text{x}) \leftrightarrow \int_0^x \frac{2}{\sqrt{\pi}} e^{-x^2} dx$$

$$\text{erfc}(\text{x}) \leftrightarrow \int_x^\infty \frac{2}{\sqrt{\pi}} e^{-x^2} dx$$

Compare to (21.3). The returned value in the example is identical to what is found in standard probability tables for $P(X < 1)$.

There are just two pre-defined distributions in MATLAB: the normal and uniform ones. However, with a simple trick and a few calculations, most other ones are easily simulated from the uniform one. The idea is that the distribution function $F(X)$, seen as a stochastic variable, is uniformly distributed,

$$F(X) = \int_{-\infty}^X f(z)dz \in U[0,1].$$

If we take a $U \in U[0,1]$ at random, we thus have that
$$X = F^{-1}(U)$$

belongs to the distribution for X. For example, take the exponential distribution with mean value $1/a$

$$f(x) = ae^{-ax} \Rightarrow$$

$$F(x) = \int_0^x ae^{-az} dz$$

$$= 1 - e^{-ax} = u \Rightarrow$$

$$x = \frac{-1}{a} \log(1 - u).$$

Here we use `histc` to produce the histogram, since it is more natural to specify the bin edges by the vector x, rather than the bin centers as done in `hist`.

```
>> ind=find(x==1);
>> F(ind)
ans =
      0.8521
>> 0.5+0.5*erf(1.0/sqrt(2))
ans =
      0.8413
>> 1-0.5*erfc(1/sqrt(2))
ans =
      0.8413
```

```
>> N=1000;
>> a=2;
>> u=rand(N,1);
>> X=-1/a*log(1-u);
>> mean(X)
ans =
      0.5173
>> x=0:0.1:max(X);
>> [fhat,xdum]=histc(X,x);
>> bar(x,fhat/sum(fhat)/0.1,'w')
>> hold on
>> plot(x,a*exp(-x*a))
>> hold off
```

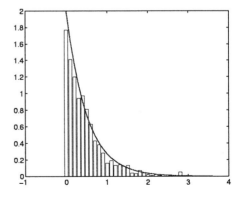

Consider now the coin tossing experiment famous in statistical theory. Suppose we assign a random variable the value +1 for head and −1 for tail. We can then simulate 10000 experiments using the uniform distribution **rand**, which generates random number between 0 and 1 with equal probability. These samples are rounded and scaled so that the result is ±1.

The cumulative sum of these samples is what is called a random walk. Suppose that you take one step forwards when the head comes up and one step backwards otherwise. Some people believe that you will end up where you start if you toss the coin many times. This is not true, although the mean value of y tends to 0. In fact, the probability that you will be say less than 100 meters from the starting point tends to zero as the number of tosses increases.

There is a celebrated result that gives a precise interval where you will be after a large number N tosses "almost surely" (which means with a probability that tends to 1 when N increases). The last statement means that you must have extremely, not to say impossible, bad luck not to be there. This result is called the *law of iterated logarithm*. The figure shows these bounds. Remember that the random walk might be outside the bounds in the beginning but will in the long run end up inside.

```
>> y=-1+2*round(rand(10000,1));
```

```
>> plot(cumsum(y))
```

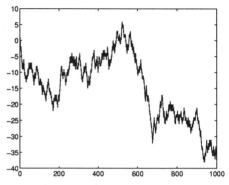

```
>> n=(3:10000)';
>> hold on
>> plot(n,...
   sqrt(0.5*n.*log(log(n))))
>> plot(n,...
   -sqrt(0.5*n.*log(log(n))))
```

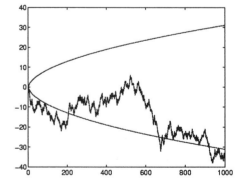

Many classical probability problems can be solved numerically by extensive simulations. One example was the coin tossing above. Other ones involve tossing a dice. We here look at a problem in chess. Suppose a lonely king starts at square A1 (lower left corner), and moves around randomly. Typical questions for this problem are:

- How often does it pass square H8? The answer computed to the right is about 0.7% of the time.

- What is the probability density function over the chess board?

- What is the mean time to go from square A1 to square H8?

We compute this numerically by simulating N moves. The code first determines all legal moves at the current position, and then takes one move at random. The function pcolor is used to illustrate the density function.

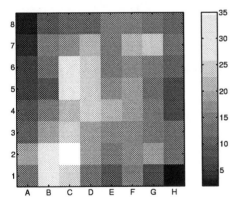

File name: `chess.m`

```
function chess
% Random king walk
moves = [
  -1 -1; -1  0; -1  1;  0  1;
   1  1;  1  0;  1 -1;  0 -1];
x(1,:)=[1 1];
A=zeros(8,8); A(1,1)=1;
N=1e3;
for i=1:N-1;
  ind = ceil(8*rand(1,1));
  x(i+1,:)=x(i,:)+moves(ind,:);
  while x(i+1,1)<1|x(i+1,1)>8|...
        x(i+1,2)<1|x(i+1,2)>8
    ind = ceil(8*rand(1,1));
    x(i+1,:)=x(i,:)+moves(ind,:);
  end
  A(x(i+1,1),x(i+1,2)) = ...
      A(x(i+1,1),x(i+1,2))+1;
end
plotfix
imagesc(A)
set(gca,'YDir','normal')
set(gca,'YTick',...
    [1 2 3 4 5 6 7 8])
set(gca,'XTick',...
    [1 2 3 4 5 6 7 8])
set(gca,'XTickLabel','A','B',...
'C','D','E','F','G','H')
colorbar
axis('square')
colormap('pink')
A
A(8,8)/N
```

22 Control Theory and the LTI Object

Linear time-invariant (LTI) systems play a central role in engineering. In MATLAB 5.3 the LTI object was introduced, which greatly decreased the number of functions by overloading taylored functions in different toolboxes to existing ones, and helped applied engineers to appreciate more advanced functions. Further, the three most well spread MATLAB toolboxes (Signal Processing, Control System, System Identification) all support the LTI object, motivating a rather detailed summary here.

This section also serves as an application example of a quite complex MATLAB object. Actually, the object **rec**, studied in guided tour 33, has much in common with the LTI object, and the recursion is a special case of linear time-invariant systems. Thus, recapitulation of Section 12 is recommended at this instance.

The control problem is to compute an appropriate control signal u to a system whose output y is observed. The goal is to get an output y that is as close as possible to a reference value r provided externally. Think of the steering servo in a car. The block diagram in Figure 22.2 uses the basic feedback structure, where a controller observes the deviation of y from the reference r and calculates the control signal u. The controller is itself an LTI system.

The basic functions generating and operating on LTI objects are summarized in Table D.

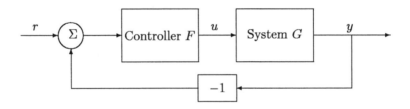

Figure 22.2 Control structure

A discrete time linear time-invariant system relating the input $u[k]$ to the output $y[k]$ is defined either as a state space model

$$x[k+1] = Ax[k] + Bu[k] \tag{22.4a}$$
$$y[k] = Cx[k] + Du[k] \tag{22.4b}$$

or as a transfer function using the $\mathcal{Z}$ transform

$$Y(z) = \frac{B(z)}{A(z)}U(z). \tag{22.5}$$

Here $B(z)$ and $A(z)$ are polynomials in z, and in both cases an implicit sampling interval T is used.

In continuous time, the following counterparts are the state space model

$$\dot{x}(t) = Ax(t) + Bu(t) \tag{22.6a}$$
$$y(t) = Cx(t) + Du(t) \tag{22.6b}$$

or as a transfer function using the Laplace (s) transform

$$Y(s) = \frac{B(s)}{A(s)}U(s). \tag{22.7}$$

The LTI object is very powerful for many purposes, and the guided tour below describes the basic ideas of system analysis and control design as summarized below:

- Creation on an LTI model **g**.

- Conversion between system representations of the continuous and discrete time models (22.4), (22.5), (22.6) or (22.7). For example **gd=ss(gc,1);** computes (22.4) from (22.6) or (22.7) (depends on how **gc** was defined).

- Design of a feedback controller F in Figure 22.2.

- Computation of the closed loop system $y = G_c r$ in Figure 22.2 using the feedback principle $u = F(r - y)$ as **gc=feedback(f*g,-1)**.

- Simulation of how y responds to step changes in r with **step(g)**.

- System identification of the LTI system G from observations of y and u. For example, **g=ss(oe([y u],[2 2 1]))**.

We will return to the damped spring studied in Sections 11.3 and 12. We will treat discrete and continuous time models in parallel, to highlight one the main strengths on the LTI object: most functions are independent on how the system was generated and how it is represented.

The system can be defined as a transfer function in two ways: by defining vectors containing the polynomial coefficients of $B(z)$ and $A(z)$, or by defining the z operator as an LTI object, for which the usual numerical operations are overloaded, and a LTI object is obtained.

```
>> a=[1 -0.9 0.81];
>> b=[0 0 1];
>> gd=tf(b,a,1)
Transfer function:
        1
-------------------
z^2 - 0.9 z + 0.81
Sampling time: 1

>> z=tf('z',1);
>> gd=1/(z^2-0.9*z+0.81);
```

The alternative is to define a state space model directly through the matrices A, B, C, D.

Transformations between these models is readily done by **gd=tf(gd)** and **gd=ss(gd)**. Notice the nicely formatted printout. For comparison, we have the following representations:

$$G(z) = \frac{1}{z^2 - 0.9z + 0.81}$$

and

$$x(t+1) = \begin{pmatrix} 0.9 & -0.81 \\ 1 & 0 \end{pmatrix} x(t) + \begin{pmatrix} 1 \\ 0 \end{pmatrix} u(t)$$

$$y(t) = \begin{pmatrix} 0 & 1 \end{pmatrix} x(t) + 0u(t)$$

The coefficients are taken from Equation (11.1).

The continuous time system that corresponds to our sampled spring model can be computed by d2c (should be pronounced 'Discrete Two Continuous'). The result is

$$G(s) = \frac{-0.6963s + 1.217}{s^2 + 0.2107s + 1.108}$$

that is (verify by doing **ss(gc)**)

$$\dot{x}(t) = \begin{pmatrix} -0.2107 & -1.108 \\ 1 & 0 \end{pmatrix} x(t) + \begin{pmatrix} 1 \\ 0 \end{pmatrix} u(t)$$

$$y(t) = \begin{pmatrix} -0.6963 & 1.217 \end{pmatrix} x(t) + 0u(t)$$

Sampling of a continuous time system is achieved with c2d as **gd=c2d(gc,1)**, and a change of the sampling rate is achieved with d2d.

```
>> A=[0.9 -0.81;1 0]; B=[1;0];
>> C=[0 1];              D=0;
>> gd=ss(A,B,C,D,1)
a =
              x1        x2
     x1       0.9     -0.81
     x2        1         0
b =
              u1
     x1        1
     x2        0
c =
              x1    x2
     y1        0     1
d =
              u1
     y1        0
Sampling time: 1
Discrete-time model.

>> gd=tf(gd);
gc=d2c(gd)

Transfer function:
   -0.6963 s + 1.217
   --------------------
   s^2 + 0.2107 s + 1.108
```

The impulse and step responses, respectively, are simulated up to time index $k = 30$. The plots below show the result for simulation of the continuous and discrete time systems, respectively. Note that interpolated horizontal lines are added between the simulation points for the discrete time simulations. If the system here represents the controller which is implemented in a computer, this is realistic. Otherwise, if the system is a sampled version of a physical system, the inter-sample behavior is unknown.

```
>> figure(1)
>> subplot(211), impulse(gc,30)
>> subplot(212), step(gc,30)

>> figure(2)
>> subplot(211),
>> impulse(gd,'o-',30)
>> subplot(212),
>> step(gd,'o-',30)
```

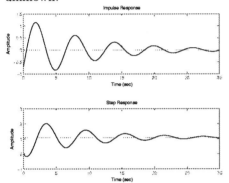

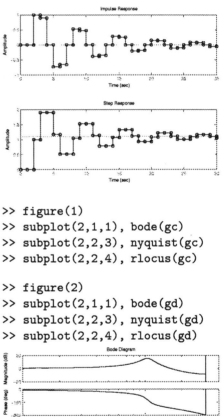

The system is illustrated in the frequency domain by a Bode diagram and a Nyquist plot. For control design, the root locus is plotted.

```
>> figure(1)
>> subplot(2,1,1), bode(gc)
>> subplot(2,2,3), nyquist(gc)
>> subplot(2,2,4), rlocus(gc)

>> figure(2)
>> subplot(2,1,1), bode(gd)
>> subplot(2,2,3), nyquist(gd)
>> subplot(2,2,4), rlocus(gd)
```

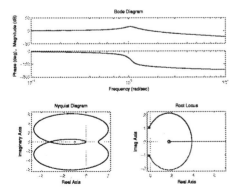

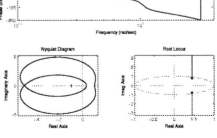

Think of the step response as what happens to your car when you hit a speed hump with a broken damper. The spring in our model is a poorly damped system. Feedback control can improve the dynamical behavior. Here we choose a method based on a state feedback controller $u(t) = r(t) - Kx(t)$, which is plugged into the state space model (22.6) or (22.4), respectively. Here $r(t)$ is a reference signal to the system, whose value the output $y(t)$ aims at following. The poles are chosen to be in -0.5 and -0.4, respectively. The step and impulse responses look better damped than the original damped spring.

The closed loop systems are given by the state space models

$$\dot{x}(t) = \begin{pmatrix} -0.9 & -0.2 \\ 1 & 0 \end{pmatrix} x(t) + \begin{pmatrix} 1 \\ 0 \end{pmatrix} u(t)$$

$$y(t) = \begin{pmatrix} -0.6963 & 1.217 \end{pmatrix} x(t) + 0u(t)$$

and

$$x(t+1) = \begin{pmatrix} 0.9 & -0.2 \\ 1 & 0 \end{pmatrix} x(t) + \begin{pmatrix} 1 \\ 0 \end{pmatrix} u(t)$$

$$y(t) = \begin{pmatrix} 0 & 1 \end{pmatrix} x(t) + 0u(t)$$

respectively. The step and impulse responses are shown below. They are better damped than for the open loop system.

```
gc=ss(gc);
kc=place(gc.a,gc.b,[-0.5 -0.4])
kc =
    0.6893    -0.9077
gcc=ss(gc.a-gc.b*kc,...
       gc.b,gc.c,gc.d);
subplot(2,1,1), step(gcc,30)
subplot(2,1,2), impulse(gcc,30)

gd=ss(gd);
kd=place(gd.a,gd.b,[0.5 0.4])
kd =
    0.0000    -0.6100
gdc=ss(gd.a-gd.b*kd,...
       gd.b,gd.c,gd.d,1);
subplot(211), step(gdc,'o-',30)
subplot(212), impulse(gdc,'o-',30)
```

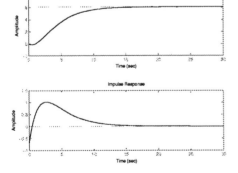

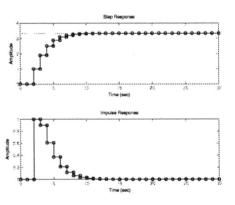

Simulation of an LTI system is obtained by lsim. Here a general input u can be used. Both **step** and **impulse** are special cases of lsim for the particular inputs

```
u=ones(N,1);
```

and

```
u=[1;zeros(N-1,1)];
```

respectively.

```
u=[1;zeros(10,1);...
    ones(10,1);zeros(20,1)];
y=lsim(gd,u);
subplot(2,1,1), stem(u)
ylabel('Input u')
subplot(2,1,2), stem(y)
ylabel('Output y')
```

The system identification toolbox contains various functions for computing an LTI system approximating the real system based on only measured input output data [y u]. Most algorithms in this field apply to discrete time data. The example here recovers the spring model based on the simulation above.

System identification assumes stochastic measurement disturbances. To avoid annoying warnings, we add a little Gaussian noise to y. Alternative functions that should give the same answer are

```
ghat=arx([y u],[2 2 1]);
ghat=armax([y u],[2 2 1 1]);
ghat=bj([y u],[2 1 0 2 1]);
```

```
y=y+0.01*randn(size(y));
gdhat=oe([y u],[2 2 1]);
gdhat=tf(gdhat)
```

```
Transfer function "u1" to "y1":
    0.003378 z + 0.9964
   ----------------------
   z^2 - 0.9008 z + 0.8098


Transfer function "v1" to "y1":
0.01155
Sampling time: 1
```

23 Dynamical Simulation with SIMULINK

Contents: Illustration of create and simulate a system in SIMULINK
Functions:
`simulink`

SIMULINK features a graphical interface to numerical integration routines, where a lot of standard operations both on discrete and continuous time signals are implemented as blocks. The user picks out blocks from standard libraries (so called *blocksets*) into an application and connects them by arrows indicating the signal flow. The user can add personal blocks by converting m-files or functions written in C or Fortran. At this stage, the standard SIMULINK blocksets listed in 27.5 should be taken a look at. SIMULINK interacts with MATLAB in the following way:

- An m-file can be incorporated in SIMULINK by using the *S-function* interface. Functions written in C or Fortran are included in the same way. Check the online help. The template file **timestwo** might be a good way to get an orientation of how an S-function is constructed.

- An LTI object can be included as a block by opening the LTI block in the block **Blocksets and Toolboxes** in the lower left corner of the main SIMULINK block in Table 27.5.

- A Simulink block diagram may be simulated from inside MATLAB by using **sim**.

SIMULINK has gained a particular impact in industry for simulating dynamical systems and designing control systems. As an example, the automotive industry considers SIMULINK to be a kind of *de-facto* standard for product documentation. For example gear boxes are delivered with a simulation model in SIMULINK which the company directly incorporates in its simulation model of the whole drive line.

There are many reasons for SIMULINK's success. One is the hierarchical way of programming with a well defined interface between different parts. Another one is the possibility of automated code generation. Real-Time Workshop is a Mathworks product for generating C-code for a particular SIMULINK model. Third-hand companies sell hardware including digital signal processors that can be programmed directly from SIMULINK.

Damped Spring

We will here revisit the damped spring, which in Section 22 was analyzed as an LTI object. It can be implemented in Simulink in many ways, using low level primitives, the state space model (22.4) or the transfer function (22.5).

We here choose the latter one. Start by typing **simulink** in MATLAB. Then open a new empty block diagram by **Ctrl-n**. Double click on the **Discrete** blockset, and drag the **Discrete Transfer Fcn** block to the empty window. Check Table 27.5 for an overview of the blocksets. To monitor its output, we pick an oscilloscope from the **Sinks** blockset, and also a **To Workspace** block from the same blockset, so the result will be available in MATLAB after a simulation.

To connect the blocks, use the mouse and click with the left button on one of the small arrows on the blocks, keep it down, and release on another arrow. To connect a line to an existing one, use the middle mouse button instead.

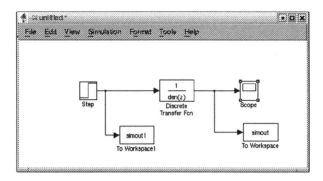

Figure 23.3 Simulink block diagram of a damped spring. A simulation is started by choosing **Start** in the **Simulation** menu, or by typing **Ctrl-t**.

To simulate a step response, we next pick the **Step** block from the **Sources** blockset. Again, we attach a **To Workspace** block, to be able to plot the result in MATLAB afterwards. The result should now look as in Figure 23.3.

We next need to define some parameters. Double click on the **Transfer function** block. The window to the left in Figure 23.4 should appear. Type the coefficients of the transfer function, just as indicated. Double click on the **Scope** to get the second window in Figure 23.4, and define the simulation time to be 30 seconds by opening the menu **Simulation-Parameters**, which opens the window to the right in Figure 23.4. Now, start the simulation by **Simulation-Start**, or by typing **Ctrl-t**.

It is possible to animate the outputs. Check the demo `onecart` in Simulink, which gives the block diagram and animation in Figure 23.5.

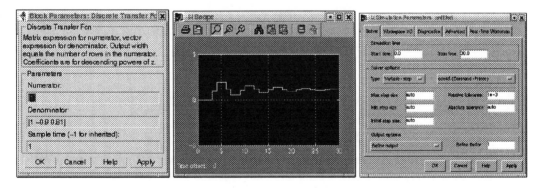

Figure 23.4 Windows that appear when double clicking on the transfer function and Scope blocks, respectively, and the simulation parameters found in the **Simulation** menu.

Power Control in Cellular Radio

We will here illustrate SIMULINK on a simple but quite realistic evaluation of power control in cellular mobile telephony. To create a new block, type `simulink` and choose a new model in the window that appears. SIMULINK 3 features a simplified block search mimicking what a file explorer looks like. Earlier versions use a large set of windows with collections of

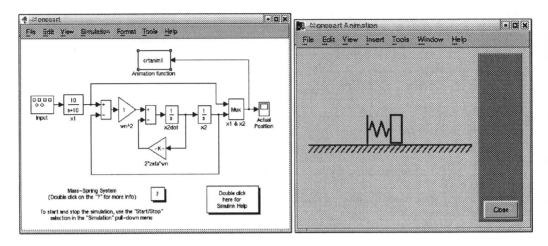

Figure 23.5 A Simulink demo containing a small animation of a damped spring is started by typing onecart in MATLAB. The right plot shows a snapshot of the animation.

blocks. After finding the blocks of interest, they are copied to the working area by common cut and paste, or drag and drop.

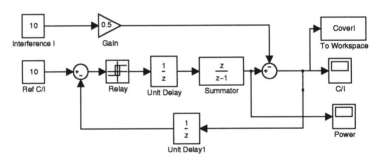

Figure 23.6 Block diagram of a power control algorithm in Simulink.

The block diagram above works as follows:

1. The goal in a digital communication system is to achieve a so called carrier to interference ratio (C/I) that is larger than a certain reference value (here 10). This threshold determines the minimum C/I for good sound quality, a smaller carrier to interference ratio results in loss of communication. The carrier power (C) is proportional to the transmitted power. The interference (I) comes from other users using the same channel. The scale of C/I is in decibel (*e.g* logarithmic).

2. The controller computes the difference between the reference value and the actual measured value of C/I. If the difference is positive, the power should be increased to achieve acceptable quality. If the difference is negative, the power should be decreased to save energy and to minimize the interference with respect to other users.

3. The protocol standardized for the so-called third generation systems assigns one bit

only for power control. This implies that the controller can only tell the mobile to either increase or decrease the power by a predefined fixed amount at each time step. The relay block has this feature.

4. The power control command is computed at the receiver (where C/I is measured), but is effectuated at the transmitter. The time delay for transmission of the power control information is modeled by a unit time delay.

5. The sum block is needed since the control commands only indicate if the control level should be increased or decreased.

6. The interference is in the depicted case another user having fixed power 10 but attenuated 0.5 due to being farther away.

7. There is also a time delay of one unit before the transmitter is affected by the new power level.

8. The resulting simulated power level is plotted in a scope block, while the C/I is displayed in a scope and exported to workspace.

When the simulation model is finished, the simulation time interval is chosen and a simulation started with **Ctrl-t** (or from the menu). The result is shown below. There is an oscillation in C/I of three quantization levels.

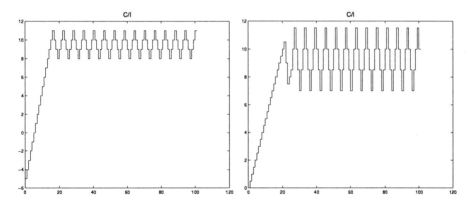

Figure 23.7 Simulation result of C/I for a user for first a single power control algorithm and then for two algorithms interfering with eachother.

Now, let's demonstrate the grouping idea. The control loop for one single user will be summarized into one block, enabling larger and more realistic simulations where the interaction between different users is examined. The SIMULINK scheme below is equivalent to the one before.

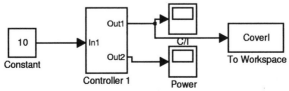

Now it is simple to connect two users.

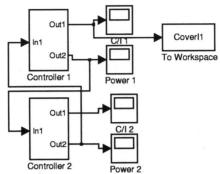

The C/I value for the first users reveal that the dynamic interactions between the users is destructive. The oscillations in C/I may be fatal for the connection, since C/I is well below the threshold a large part of the time.

24 Ordinary Differential Equations

> **Contents:** Simulation of ODE's
> **Functions:**
> ode45

The core of SIMULINK consists of advanced methods for simulating an *ordinary differential equation* (ODE):

$$\dot{x}(t) = f(x(t)).$$

The solution in general can be written

$$x(t) = x(0) + \int_0^t f(x(s))ds.$$

However, the integral has only an explicit solution in simple examples. For instance,

$$\dot{x}(t) = -ax(t) \Rightarrow x(t) = x(0)e^{-at}.$$

In the general case, when $f(x)$ is non-linear and when $x(t)$ is vector valued, numerical methods have to be used. Essentially, these evaluate the integral by summing up the integrand over small time intervals,

$$x(t) = x(t - h) + \int_{t-h}^t f(x(s))ds$$
$$\approx x(t - h) + hf(x(t)).$$

The size h can be fixed, but more often it is determined adaptively, and iterative refining methods are applied to improve the accuracy. Many of the low-level functions are available in MATLAB, and we will here in particular use one called **ode45**. The example below can also be simulated in SIMULINK of course.

A guided tour 58 (Bouncing ball)
Many standard functions in Matlab are good examples on the use of numerical methods. As an example of an application, we consider simulation of an ordinary differential equation. We will consider the motion of a bouncing ball under influence of air drag. Denote the absolute velocity v and its direction φ. The air drag is proportional to the squared velocity

$$F_a = cv^2$$

and directed opposite the velocity vector. The other force influencing the ball is the gravity,

$$F_g = mg$$

which is directed downwards.

That is, in the x–y coordinate system, the differential equations of motion can be written, using $\sum F = m\dot{v}$,

$$\dot{v}_x = -\frac{cv^2}{m}\cos(\varphi)$$

$$\dot{v}_y = -\frac{cv^2}{m}\sin(\varphi) - g$$

where

$$v^2 = v_x^2 + v_y^2$$

$$\varphi = \arctan\frac{v_y}{v_x}$$

File name: `odefun.m`

```
function Xdot=odefun(t,X)
% define the ODE
c=0.1; m=1; g=1;
x=X(1); y=X(2); vx=X(3); vy=X(4)

g=sign(y)*g;
phi=atan2(vy,vx);
Fa=c*(vx^2+vy^2);
vxdot=-Fa/m*cos(phi);
vydot=-g-Fa/m*sin(phi);
xdot=vx;
ydot=vy;

Xdot=[xdot,ydot,vxdot,vydot]';
```

We are also interested in the ball's location x and y. We can write their differential equations in terms of the velocity as

$$\dot{x} = v_x$$

$$\dot{y} = v_y$$

The variables x, y, v_x and v_y are called the state of the system. The m-file for computing the state derivatives shown to the right. A small trick is used to simulate the bounces. The sign of the gravitational force is changed when the ball passes $y = 0$ rather than changing the sign of v_y. The absolute value of y gives correct values of the bounces.

Now the simulation of the ball is easy. We define the initial state as $(x, y) = (0, 1)$ and $(v_x, v_y) = (1, 1)$ and simulate the motion from time 0 to 10 seconds.

```
>> X0=[0,1,1,1]';
>> [t1,X]=ode45('odefun',0,10,X0);
```

We can plot the evaluation of the position as a function of time, as shown to the left below. Note the the time axis delivered by Matlab is not equidistant but rather optimized for fast computation.

```
>> plot(t1,X(:,1:2))
>> plot(X(:,1),abs(X(:,2)))
```

Even more revealing is the x–y plot to the right below.

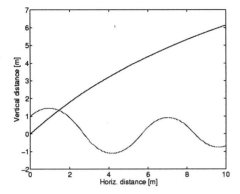

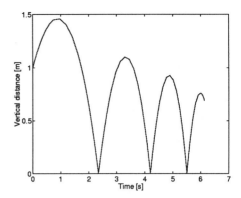

If we want an equidistant time axis, we have to use a loop. The example script-file to the right can be used. Alternatively you can use a fixed-step solver with the syntax

`ode45('odefun',[0:0.5.10],x0);`

instead of the loop shown to the right.

```
X0=[0,1,1,1];
M=1;
t=0:.5:10;
for i=2:length(t);
  [dum,X0]=ode45('odefun',...
          t(i-1),t(i),X0(M,:)');
  [M,dum]=size(X0);
  x(i-1)=X0(M,1);
  y(i-1)=X0(M,2);
  vx(i-1)=X0(M,3);
  vy(i-1)=X0(M,4);
end;
>> plot(x,abs(y),'o')
>> hold on
>> quiver(x,abs(y),...
          vx,vy.*sign(y),0.5)
>> hold off
```

We can now plot an 'o' at the ball's position every 0.5 second. Furthermore, we can add the velocity vector at each point using the `quiver` function.

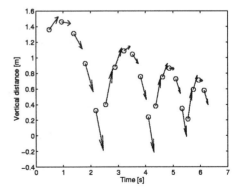

25   Signal processing

Contents: Illustration of signal and image processing, the Fourier transform and filtering
Functions:
`filter, conv, fft, ifft, fft2, ifft2, freqz, abs, angle`
Sound facilities:

`sound, soundsc, wavread, wavewrite, wavplay, wavrecord, auread, auwrite`

Sound examples:

`gong, handel, laughter, train, chirp`

Image facilities:

`imread, imwrite`

Image examples:

`earth, penny, gatlin, gatlin2, detail, flujet, spine, clown`

We will illustrate three basic signal processing tasks in separate guided tours:

- Sampling of a chirp and reconstruction of a sinusoid.

- To find a sinusoid signal hidden in noise.

- Image processing using standard images available in MATLAB.

The purpose is to introduce the principles of signal processing with illustrative examples using sound (MATLAB vectors) and images (MATLAB matrices). Little mathematical background knowledge is required here, in contrast to signal processing courses which heavily rely on transform theory, linear algebra and complex functions. The point with the chosen examples is of course to support conventional graphics with sound and images where the different phenomena can be more easily appreciated.

The guided tours will make use of the *Fast Fourier Transform* `fft` to compute the *Discrete Fourier Transform* (DFT) of a sampled signal. In an algebraic interpretation, the DFT is a mapping from one vector to another $Y = DFT(y)$, and the MATLAB syntax is simply `Y=fft(y);`, where `size(Y)` is equal to `size(y)`. The guided tours will illustrate some of the benefits of this mapping.

The first tour requires sound facilities on your platform to really be instructive. PC users will probably have no problems. The reader is encouraged to import sound and images from WWW or existing hardware like scanners and microphone. Windows users can for instance use `wavrecord` in which MATLAB takes data from the microphone directly, without the need for any external program or data format.

A guided tour 59 (Reconstruction)

Let us start with a simple example to test the sound on your computer. The demo signal train is loaded. The workspace now contains the vector y and a scalar *sampling frequency* Fs. The platform independent function sound takes a vector and a sampling frequency (default is 8192 Hz) as inputs, and uses the vector to produce a sound (train whistle in this case). On some platforms (Sun), the sampling frequency is fixed to 8192 Hz, so the second argument has no meaning.

```
>> load train
>> whos
   Name        Size            Bytes

   Fs          1x1                 8
   y           12880x1        103040

Grand total is 12881 elements
using 103048 bytes
>> sound(y,Fs)
```

Sound is a continuous air pressure wave, while the input is a vector of real numbers. How can a vector sound? Apparently, the vector needs to be converted to a continuous signal, before it is (amplified and) sent to the loudspeaker.

The demo signals in MATLAB as well as all digital signals are obtained from sampling. That is, the signal is read off at regular time intervals. We will start with considering the *reconstruction* problem: how is the continuous signal computed from the samples?

A sinusoid with frequency 128 Hz is generated, *sampled* with frequency 8192 Hz, which means that 8192 values per second are read off. This is sufficiently many for the plot of the first 25 ms to look like a smooth continuous curve, and sound produces a clear tone. Remember that the plot function interpolates a straight line between each two vector elements.

```
>> f0=128; %Signal frequency [Hz]
>> Fs=8192;%Sampling frequency
>> Ts=1/Fs;%Sampling interval [s]
>> t=0:Ts:1-Ts; %1s time vector
>> y=sin(2*pi*f0*t);
>> plot(t(1:200),y(1:200))
>> sound(y,Fs)
>> pause(1)
```

The last pause is required for some combinations of platform and sound card, otherwise, when another sound output is requested before the previous one is executed, an error message is produced by MATLAB.

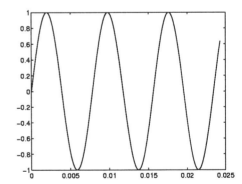

Since continuous signals cannot be illustrated exactly in MATLAB, we will use the following convention that if the sampling is sufficiently fast, it can be considered as a continuous time signal.

- Fast sampling (8192 Hz) is used to approximate the continuous curve. The approximation is good both by visual and audio inspection in plots (on a reasonable time scale) and on many sound devices.

- A substantially slower sampling (1024 Hz or 2048 Hz) is used to illustrate the various effects of reconstruction and sampling.

The fast sampling rate is chosen so that the examples given below should work on all platforms. Figure 25.8 gives an overview of this guided tour.

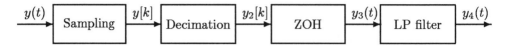

Figure 25.8 A sinusoid $y(t)$ is sampled $y[k]$ with frequency 8192Hz, which is considered to give an 'almost' continuous signal, so $y[k/8192] \approx y(t)$ both to the eye in plots and to the ears when played in a loudspeaker. The sampled signal is then decimated a factor 8, which gives $y_2[k]$. To reconstruct $y(t)$, the signal first passes a *zero-order hold* (ZOH) circuit, which holds the signal constant between the sampling instants. The continuous time signal $y_3(t)$ has a clear distortion when listening to it. After appropriate low-pass filtering, the original signal is reconstructed, so $y_4(t) = y(t)$ (at least with an ideal filter).

If we want to store the signal with sampling frequency 1024 Hz, we can pick out every 8'th sample. This *decimation* with a factor of 8 is possible without information loss (this is a consequence of a famous result: the *sampling theorem*). More on this later. In this way, the need for memory space is reduced with a factor of 8. The remaining problem is how to reconstruct the signal (here the fast sampled signal).

`plot` is used for the continuous curve, and `stem` for the sampled one.

On some platforms (*e.g.* Sun), the sampling frequency is fixed to 8192 Hz always. Otherwise, you can listen to the down-sampled signal. The *distortion* is obvious.

```
>> decfactor=8;
>> Fs2=Fs/decfactor;
>> Ts2=1/Fs2;
>> t2=t(1:8:end);
>> y2=y(1:8:end);
>> plot(t(1:200),y(1:200))
>> hold on
>> stem(t2(1:25),y2(1:25))
>> hold off
>> sound(y2,Fs2)
>> pause(1)
```

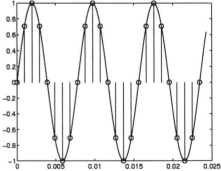

The simplest and most common first step in reconstruction is to use a so called *zero order hold* principle, where the latest sampled value is hold constant to the next one. In this way, a continuous signal reminding of a staircase is obtained. The signal sounds distorted.

The abrupt transitions introduce high-frequency components in the signal. This is the cause of the distortion. The next step in the reconstruction is to *low-pass* filter the signal. Before that, we will illustrate what the distortion is in the frequency domain.

We here use the *discrete Fourier transform* to analyze the spectral energy for each frequency in the signal. The first subplot illustrates that the original signal has energy content at only one single frequency, namely 128 Hz. The next subplot shows that the zero order hold reconstruction has a number of "unexpected" tones. This is what appears as distortion to the ear.

The harmonics that give rise to the distortion can now be removed by a low-pass filter. We will not go into the details of filter design, but just state that **butter** is one way to construct a filter, and **filter** performs the filtering operation, which is just a vector to vector transformation. Table 27.1 gives some more options.

The plot to the left below shows the appropriate filter which has a cut-off frequency of FS/2/16.

```
>> y3=kron(y2,ones(1,decfactor));
>> Fs3=decfactor*Fs2;
>> plot(t,y,'--',t,y3,'-')
>> hold on, stem(t2,y2), hold off
>> axis([0 0.01 -1 1])
>> legend('y(t)','y_3(t)',...
          'y_2[k]',3)
>> sound(y3)
```

```
>> Y=fft(y);
>> Y3=fft(y3);
>> N=length(y);
>> subplot(211)
>> semilogy(0:Fs/2,...
        abs([Y(1:N/2+1)]).^2)
>> title('Original signal')
>> subplot(212)
>> semilogy(0:Fs/2,...
        abs([Y3(1:N/2+1)]).^2)
>> title('Reconst. signal')
>> xlabel('Frequency [Hz]')
>> ylabel('Energy content')

>> [b,a]=butter(4,1/16);
```

Next, the filter is applied to the ZOH signal. The filter introduces a transient effect in the very beginning of the signal, and a time delay in stationarity (after the transient). However, neither the transient nor the time delay is audible.

```
>> y4=filter(b,a,y3);
>> plot(t,y,'--',...
        t,y3,'-',t,y4,'-.')
>> hold on
>> stem(t2,y2)
>> hold off
>> axis([0 0.01 -1 1])
>> legend('y(t)','y_3(t)',...
          'y_4(t)','y_2[k]',3)
>> sound(y4)
```

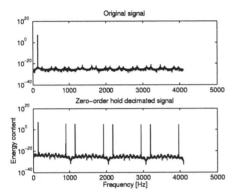

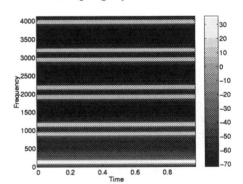

Another way to show frequency content of a signal is to plot a *spectrogram*. The MATLAB function `specgram` is quite easy to use for this purpose. The energy content is with the default color map shown on a color scale, where dark red means high energy and light blue low energy content. The function `colorbar` adds the scale to the right in the plot. On the gray scale in the figure here, the line at the bottom corresponds to the original 128 Hz signal, the other lines are *harmonics* (integer multiples of the sinusoid frequency). Having time on one axis makes it possible to study how the energy content varies with time.

```
>> specgram(y3,[],Fs)
>> colorbar
>> colormap('gray')
```

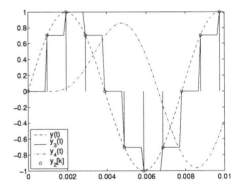

Sampling is the inverse operation of reconstruction. As before, we generate a signal with 8192 samples per second and consider this to be an approximation of a continuous signal. Sampling is then a process of picking out a subsequence of the test signal.

A guided tour 60 (Sampling)

The test signal will be a so called *chirp* considered earlier, which is defined as a tone with growing frequency. As the spectrogram reveals, the frequency content goes from zero to 16384 Hz during two seconds. The sound confirms that everything works as expected, but the sound hardware may not be capable to reproduce the very low and high frequencies. Also the human ear will attenuate high and low frequencies so much that they are inaudible.

```
>> t=1:0.25:2*8192;
>> Fs=2^15;
>> y=sin(2*pi*t.^2/t(end));
>> specgram(y,[],Fs);
>> colorbar
>> sound(y,Fs)
```

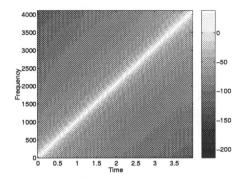

Assume we decimate this signal a factor of four in an attempt to save memory. Thus, we pick out every fourth sample.

The spectrogram now shows a funny M-pattern, where the energy content increases with frequency in the beginning, then decrease, increase and finally decrease again. Why??

The sound resembles a whistle.

```
>> decfactor=4;
>> Fs=Fs/decfactor;
>> ys=y(1:decfactor:end);
>> specgram(ys,[],Fs);
>> colorbar
>> sound(ys)
```

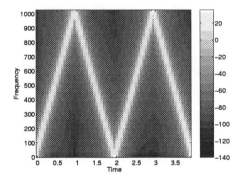

What we have seen and heard is a consequence of the *sampling theorem*: no frequency over one half of the sampling frequency appears correctly after sampling. Instead it pops up at a completely different frequency. This is called *aliasing*: a frequency component in the original signal appears under a false name after sampling. There is no way to reconstruct higher frequencies than one half of the sampling frequency. The best thing to do is to low-pass filter the signal before it is sampled and remove everything over half the sampling frequency. In this way, no false tones appear after the reconstruction due to aliasing.

To remove signal energy over half the sampling frequency, a low-pass filter is constructed which removes high-frequency energy before sampling. The M-pattern is still visible in the plot but the last part is attenuated so it is not audible anymore.

```
>> [b,a]=butter(4,1/decfactor/2);
>> yf=filter(b,a,y);
>> ys=yf(1:decfactor:end);
>> sound(ys)
>> specgram(ys,[],Fs);
>> colorbar
```

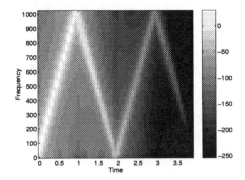

There are plenty of applications where one wants to find a sinusoid in a noisy signal and estimate its frequency, for instance finding radio carriers or radar signals. The next tour shows how filtering and Fourier transform techniques can be used for this purpose.

A guided tour 61 (Signal in noise)

We will here work with a signal that consists of a pure sinusoid with white noise (random numbers) added. We generate this signal with frequency $f = 1/8 = 0.125$ and random variables from the normal distribution are used.

```
>> t=1:128;
>> f=1/8;
>> u=1*sin(2*pi*f*t)+...
      randn(size(t));
>> plot(u)
```

We assume that the sampling interval is one, so t denotes both the vector index and absolute time.

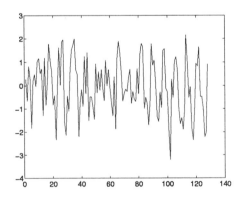

It is hard to visually detect the sinusoid. The standard tool for detecting sinusoids in a signal is the *Fourier transform*, which gives a measure of signal content for each frequency. As mentioned in earlier in this section, the method to compute this measure for a sampled signal is called the *Discrete Fourier Transform*, and the best algorithm to compute it is called *FFT* (*Fast Fourier Transform*), implemented in `fft`. The input is a vector (of real numbers usually) and the output a vector of complex numbers. One of the most complicated tasks for beginners in signal processing is to understand the symmetries and indexing in the result. In this case, we have with MATLAB indexing:

$$U(k) = U^*(N + 2 - k), \quad k = 2, 3, \ldots, N/2,$$

where * denotes complex conjugation. That means that the second part is basically a mirror of the first part, while $U(1)$ and $U(N/2+1)$ are unique. This is the case for the transform of all real valued signals, and we will not plot the mirror frequencies in the sequel.

We distinguish a peak at frequency 0.125.

An accurate value of the maximum is obtained with the max function (**ginput** can be used as well).

A classical way to extract a signal from the noise is to use filters. A simple low-pass filter that can be used is

$$H(z) = 0.25(z^3 + z^2 + z + 1).$$

The filter is defined as

$$y(t) = 0.25(u(t) + u(t-1) + u(t-2) + u(t-3))$$

for this particular $H(z)$. This filter simply takes the local average of the signal. The output from the filter is computed by convolution using the `conv` function. One can now guess that there is a pure sinusoid in the signal.

```
>> U=fft(u);
>> w=1/128*(0:64);
>> plot(w,abs(U(1:65)))
```

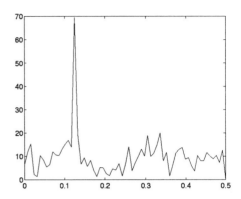

```
>> [dum,ind]=max(abs(U));
>> w(ind)
ans =
        0.1250
>> h=0.25*[1 1 1 1];
>> y=conv(u,h);
>> plot(y)
```

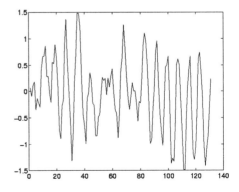

The Fourier transform reveals that high frequencies are attenuated while the low frequencies, where the sinusoid is found, are unchanged.

```
>> Y=fft(y);
>> plot(w,abs(Y(1:65)));
```

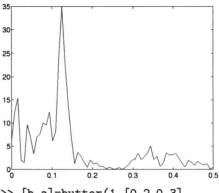

More sophisticated filters can be computed using functions in the signal processing toolbox. For instance, a first order Butterworth bandpass filter centered around the sinusoid is given by

$$H(z) = \frac{0.1367z^2 - 0.1367}{z^2 - 1.2361z + 0.7265}.$$

In the time domain, this means that

$$y(t) - 1.2361y(t-1) + 0.7265y(t-2)$$
$$= 0.1367u(t) - 0.1367u(t-2).$$

The bandpass filter is designed to attenuate all frequencies except an interval around 0.125. The filtered signal is now very similar to a pure sinusoid.

Note that *normalized frequencies* are used in filter design, where the frequency argument is always in the interval $[0, 1]$, where 1 means half the sampling frequency. In this case, $[0.2\,0.3]$ means that we want to keep all frequencies $f \in [0.10.15]$.

```
>> [b,a]=butter(1,[0.2 0.3],...
                      'bandpass')
b =
        0.1367          0    -0.1367
a =
        1.0000    -1.2361     0.7265
>> y=filter(b,a,u);
>> plot(y)
```

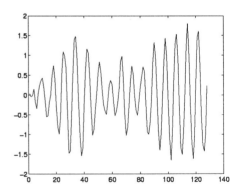

The peak in the Fourier transform is much more distinct after bandpass filtering.

```
>> Y=fft(y);
>> plot(w,abs(Y(1:65)));
```

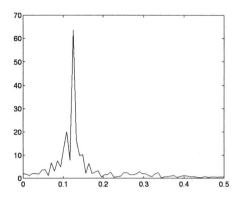

Basic filtering can be used to affect sound is different manners. We will here provide some examples.

A guided tour 62 (Funny sound effects)

We here use the laughter signal **laughter** delivered with MATLAB. It is quite long, so to make the computations below a bit faster, we truncate it. The plot reveals that the laughing people is taking a breathe before sample 14000, so let us break the signal here.

```
>> load laughter
>> plot(y)
>> sound(y)
```

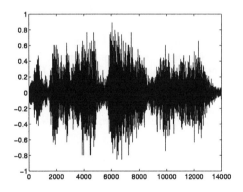

```
>> y=y(1:14000); % Truncation
>> sound(y)
```

An echo effect can be constructed by adding the sound vector to itself with a little delay included. Suppose we want a 0.5 s echo, which is amplified with a factor of 1.2. One way is to pad 4096 zeros, first at the beginning, then at the end of the vector, and add these two vectors with a scaling factor of 1 and 1.2, respectively:

```
>> yf2=[y;zeros(4096,1)]+...
       1.2*[zeros(4096,1);y];
```

An alternative method is to construct a filter which does the same thing. First, a little notation. The continuous time signal is denoted $y(t)$. After sampling we get the sequence $y[k] = y(kT)$, where T is the sampling interval. The shift operator q^{-1} shifts the signal one sample backwards, so $q^{-1}y[k] = y[k-1]$. The wanted filter operation can then be written $(1+1.2q^{-4096})y[k] = y[k]+1.2y[k-4096]$. The vector b consists of the coefficients of a filter polynomial $B(q) = b_0 + b_1q^{-1} + \cdots + b_mq^{-m}$. Thus, a lot of zeros need to be included in the vector b. The plot shows the so called *impulse response* of the filter. The interpretation is that first the signal itself is passed with time delay zero and amplification 1, then comes the signal delayed with 0.5 seconds and amplified with a factor of 1.2.

The echo effect is clearer on a speech signal. Record your own speech with **wavrecord** (PC)!

```
>> y=[y;zeros(size(y))];
>> b=[1 zeros(1,Fs/2) 1.2];
>> yf2=filter(b,1,y);
>> sound(yf2)
>> stem((0:1)*0.5,[1 1.2])
```

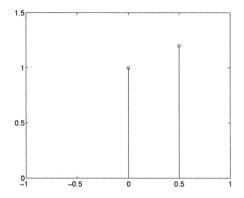

Another effect is achieved if we assume that we have two reflectors (at a distance of 340/4 meters), and that the sound bounces back and forth, each time attenuated with, say, 0.8. The impulse response illustrates the filter interpretation of this effect, which is called *reverbation*. The filter can be written

$$H(q) = \frac{1}{1 + 0.8q^{-4096}}$$
$$= 1 - 0.8q^{-4096} + 0.8^2 q^{-2 \cdot 4096} - \ldots$$

which shows that the sound signal is delayed an integer multiple of 4096 samples (half a second) and attenuated a factor of 0.8 each time.

```
>> y=[y;zeros(size(y))];
>> a=[1 zeros(1,Fs/2) 0.8];
>> stem((0:9)*0.5,0.8.^(0:9))
>> yf3=filter(1,a,y);
>> sound(yf3)
```

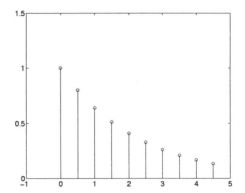

A guided tour 63 (Image processing)
The first test image is a satellite picture of the earth. The camera is located approximately over the south of Africa. The quality is poor, at least compared to what cheap scanners and digital cameras can achieve. Still, many details are visible, but zooming is not recommended. The resolution is 250 times 257 and each color is stored with an 8-bit integer, yielding a total of 256 colors.

Images of various standard formats can be imported with **imread**, and the processed image can be exported with **imwrite**. Try it out on images from WWW, or even on a picture of yourself!

```
>> load earth
>> image(X)
>> colormap(map)
```

There are a few standard filter operations for images in MATLAB. The first and second derivative of an image, seen as a two-dimensional function, are computed by `gradient` and `del2` respectively. An alternative way to approximate a gradient is to compute successive differences with `diff`.

For instance, the `del2` function approximates the second derivative at a certain matrix element with its actual value minus the average of its four neighbors. The function is here used to find the contours of the continents.

```
>> Ygrad=gradient(X);
>> Ydel=del2(X);
>> Ydiff=diff(X);
>> image(Ydel)
>> colormap(copper)
```

The second test image is a portrait of a baboon.

```
>> load mandrill
>> imagesc(X)
>> colormap(gray)
```

A common image distortion model is that noise is added to the matrix due to transmission or storing errors. Here we add artificial Gaussian noise from the random number generator. There are 256 shades in the gray scale, and the added noise operates pointwise in the matrix. The standard deviation of the noise is 20 (rounded to an integer).

```
>> Xn=round(X+20*randn(size(X)));
>> imagesc(Xn)
```

One way to improve the noisy image quality is to average neighboring points. By *convolution* using a small square matrix with equal elements that sum up to 1, each point is replaced by the neighborhood average. To the right, a 3 times 3 convolution kernel is applied. The noise is attenuated.

```
>> Xnf=round(...
      conv2(Xn,ones(3)/9));
>> imagesc(Xnf)
```

By averaging over more points, here a seven times seven matrix, the averaging effect increases. On the other hand, the resolution becomes worse. The best result is a rather subjective compromise.

```
>> Xnf=round(...
       conv2(Xn,ones(7)/49));
>> imagesc(Xnf)
```

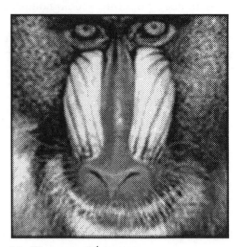

For image compression, for transmission or storing, a trivial idea is to down-sample the image. Here we try to down-sample the image a factor of 10, which means that only 1% of the memory is required. This kind of squares in the image is sometimes seen in for instance newspapers, where block transforms as JPEG have been used.

Note that a low-pass operation (here using local averaging) is performed first to avoid most of the alias effects. Compare with the section on sampling.

```
>> Xlp=round(...
       conv2(X,ones(10)/100));
>> Xdec=Xlp(1:10:end,1:10:end);
>> imagesc(Xdec)
```

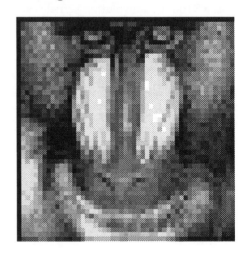

There are many standards for compressing images, like *JPEG* and *GIF*. These are based on different kind of transforms of the image, and we illustrate the redundancy that is highlighted after transformation by using the *Fourier transform* in `fft2`. Compare the quality to the previous plot, which has the same compression rate.

The code keeps only 1 % of the transform variables, which implies a compression rate of of 1 %.

```
>> load mandrill
>> Y=fftshift(fft2(X));
>> Z=zeros(size(Y));
>> Z(225:275,225:275)=...
        Y(225:275,225:275);
>> U=ifft2(fftshift(Z));
>> imagesc(real(U))
>> colormap(gray)
```

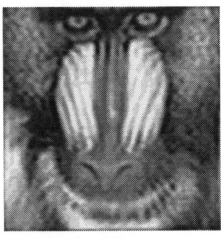

A plot of the transform reveals how little information the image contains. Only some samples in the middle contributes to the image, and this is why the masking above worked so well.

```
>> imagesc(abs(Y))
>> brighten(0.8)
```

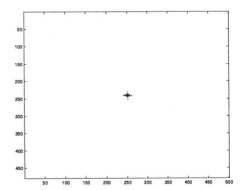

26   Communication systems

We will in this section implement simple but illustrative examples of all major components in a general communication system. The example is realistic enough to be used for communication between two computers. The information will be sent from the loudspeaker on one computer to the microphone on another computer, and the design task is to convert information to sound and then back to information in a robust and reliable way. The key performance measure of a communication system is how the *bit error rate* (BER) depends on the *signal to noise ratio* (SNR) for a given transmission rate. That is, how much interfering noise in the room can be tolerated? It is quite instructive to use an acoustic channel where one can listen to the transmitted signal and the interference.

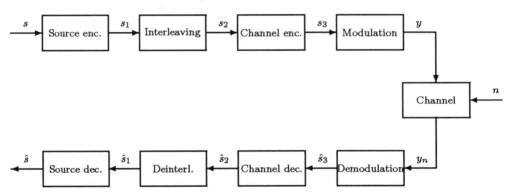

Figure 26.9 Block diagram of a general communication system, which also introduces the notation used in the guided tour.

The basic building blocks are in common with commercial standards, and a block diagram as often shown in telecommunication books is given in Figure 26.9. Briefly, the role of each step is as follows:

1. The information is here a string of ASCII characters. Alternatives include real-time applications as voice and video, where the information is split up in frames, which are similar to the letters in our example.

2. The *source encoding* digitizes the information and could also include compressions as *zip* for text, transform or model based compressors for voice and video. In our example, no compression is done.

3. *Interleaving* means that the information is shuffled around in time in a certain pattern. This is useful when transmission errors most likely occur in bursts., If we lose ten characters in a row in our example, an important word may be missing. On the other hand, if we first shuffle around the characters, the message might still be readable if the ten incorrect characters are spread out. After all, text contains a lot of redundancy. Burst errors occurs in several voice and music applications, so interleaving is an important step in cellular radio systems and CD players, for instance. Interleaving can also be used for *encryption*, since the receiver must know the scrambling pattern exactly, otherwise the message would be illegible.

4. The idea in *channel coding* is to add redundant information to the transmitted message, which in the receiver is used for error detection and error correction. This step is not included in our example.

5. The task of *modulation* is to adapt the digital information to the analog channel. We use a carrier in the audible frequency interval, so that the loudspeaker and microphone can be used as the receiver and transmitter, respectively.

A guided tour 64 (A communication system)

The text message is first converted to a vector of ASCII code using `double`. Each ASCII code word is an integer between 0 and 255, and the corresponding eighth bit binary number is computed with `dec2bin`. The result is a string, *e.g.* '01010100', for each element in the vector. Here `tmp2` is a matrix with 8 rows and `length(s)` columns. `tmp2(:)` makes one long vector of this, which can then be converted to a vector of real numbers with `str2num`.

Source decoding is the inverse operation.

To scramble the bits in the message, we can use the random number generator to compute a random index list. `ind` will contain all integers between 1 and `length(s1)` in a random order. If the receiver has the same random number generator and knows the 'code word', here 1, it can compute the same list and unscramble the bit order.

Error control by channel coding is one of the most critical steps in a communication system. For simplicity, we leave this out here.

```
>> s='Telecommunication!';
>> % Source coding
>> tmp1=double(s);
>> tmp2=dec2bin(tmp1,8)';
>> s1=str2num(tmp2(:));

>> % Source decoding
>> tmp1=num2str(s1hat);
>> tmp2=reshape(tmp1,8,...
        length(tmp1)/8)';
>> tmp3=bin2dec(tmp2);
>> shat=char(tmp3');

>> % Interleaving
>> rand('seed',1);
>> [dum,ind]=sort(rand(size(s1)));
>> s2=s1(ind);
>>
>> % Deinterleaving
>> rand('seed',1)
>> [dum,ind]=sort(rand(...
        size(s2hat)));
>> s1hat(ind)=s2hat;
>> s1hat=s1hat';
>> % Channel coding
>> s3=s2;
```

Key design parameters for capacity calculations in the modulation step are:

- The *bit rate* f_{bit}.

- The character rate will for us be $f_{bit}/8$. Note however, that if only letters are allowed in the message, we can assign fewer bits to each character. This is a design choice in source coding.

- The *symbol rate* f_{symbol} depends on how many bits we want to put on the carrier at each time. Here we will use a binary phase shift, where one bit decides the phase of the carrier, which means that $f_{symbol} = f_{bit}$.

- The carrier frequency $f_{carrier}$. The number of periods of the carrier per symbol, $L = f_{carrier}/f_{symbol}$, is here 10 only. A larger number means higher robustness to noise.

- Since we here compute a sampled version of the continuous time carrier, we need to design the over-sampling frequency $f_{sampling}$. The analog signal is then generated by the audio card, compare to guided tour 59. It is here chosen as $f_{sampling} = K f_{carrier}$.

The phase shifted modulated signal can be written as

$$y(t) = 2(s_4(\lfloor t \cdot f_{sampling} \rfloor) - 0.5) \cdot \sin(2\pi f_{carrier} t),$$

where $\lfloor x \rfloor$ denotes the integer part. To keep the signal band-limited, abrupt changes in the derivative of $y(t)$, which occur each time s_4 changes, should be avoided. The remedy is *pulse shaping*, where each symbol period is multiplied with a pulse. With the chosen pulse, all derivatives at the border of each symbol period are exactly zero, implying a smooth signal. The code generates the carrier during one symbol period, and `kron` is used to put these together.

```
>> % Modulation
>> M=length(s3); % Message size
>> L=10;% Carrier intervals/symbol
>> K=40;% Samples/carrier interval
>> N=M*L*K;% Total number of data
>> fc=2500;% Carrier frequency
>> % Symbol rate = fc/L
>> % Sampling interval = fc*K
>> % Carrier for one symbol
>> tmp1=sin((1:K*L)'*2*pi/K);
>> % Pulse shaping
>> p=sin((1:K*L)'*pi/K/L);
>> tmp2=tmp1.*p;
>> y=kron(s3,tmp2)-...
     kron(1-s3,tmp2);
>>
>> % Demodulation
>> tmp1=sin((1:K*L)'*2*pi/K);
>> p=sin((1:K*L)'*pi/K/L);
>> tmp2=tmp1.*p;
>> yone=kron(ones(M,1),tmp2);
>> ycorr1=yone.*yn;
>> ycorr2=reshape(ycorr1,L*K,M);
>> s2hat=0.5*...
        (1+sign(mean(ycorr2)));
```

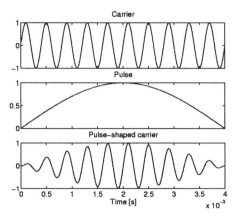

The channel model is the simplest possible; Gaussian distributed noise is added to the real-valued signal, and the noise scaling is chosen to give the specified signal to noise ratio.

```
>> % Channel
>> yn=y+sqrt(10^(-snr/10)/...
   (y'*y/length(y)))*randn(size(y));
```

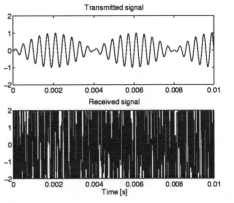

We now put everything together according to Figure 26.9, and collect the code above in a function called **comsys**. The call below specifies SNR to be -6dB, which means that the noise power is approximately four times higher than the signal power. For a particular realization, this results in two bit errors, here affecting two different characters.

```
>> comsys(-6)
s =
Telecommunication!
s1 =   01010100011001010110110001100101...
s2 =   11010100011001010101101000101000...
s1hat=01010100011001010110110001100101...
shat =
Telecoomuĭication!
ber =
    0.0132
ler =
    0.1053
```

File name: `comsys.m`

```
function [ber,ler]=comsys(snr)
if nargin<1; snr=0; end;

% Message
s='Telecommunication!';
% Source encoding
% Interleaving
% Channel encoding
% Modulation
% Channel
% Demodulation
% Deinterleaving
% Channel decoding
% Source decoding

ber=length(find(s1hat-s1~=0))/...
            length(s1);
ler=length(find(shat-s~=0))/...
            length(s);
if nargout==0
  disp(['s1hat = ',...
        num2str(s1hat(1:32))'])
  s,shat,ber,ler
end
```

The example above gave a BER of 1.3% for -6dB SNR. It would be interesting to compute a realistic function of how BER depends on SNR in *the average*. For this purpose, we use *Monte Carlo* simulations, where the same message is sent many times, and the mean and standard deviation for SNR is computed. By varying the SNR, we get the figure below. We here also plot the letter error rate (LER). After all, the person who is reading the received message is only interesting in letters, and not the individual bits, so such high-level error measures are relevant in practice.

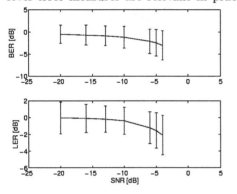

```
n=10;
snr=[10 5 0 -5 -10 -15 -20];
m=length(snr);
for i=1:n;
 for j=1:m
  [ber(i,j),ler(i,j)]=...
                comsys(snr(j));
 end
 waitbar(i/n)
end
figure
subplot(2,1,1)
errorbar(snr,log10(mean(ber)),...
        log10(std(ber)/sqrt(n)))
ylabel('BER [dB]')
subplot(2,1,2)
errorbar(snr,log10(mean(ler)),...
        log10(std(ler)/sqrt(n)))
ylabel('LER [dB]')
xlabel('SNR [dB]')
```

27 Documentation, presentation and animation

> **Contents:** Documentation using HTML, LaTeX, MS Word and other word processors.
> Tools suitable for presentations. Animations.
> **Functions:**
> print, notebook, comet, comet3, movie, getframe, movie2avi

Using MATLAB professionally in engineering applications requires the ability to document the work both in printed report form, but perhaps also in live presentation form. The principles for documenting your work performed in MATLAB are quite simple but yet very powerful:

- Figures are saved as Encapsulated PostScript files using `print -deps file.eps`.
  Change this to `print -depsc file.eps` if you want the figure in color. For users of *Microsoft office* products, the option `print -dmeta -file.wmf` can be used to export the file to the clipboard. The same thing is achieved with `print -dbitmap file.bmp`, sending a fixed resolution bitmap to the clipboard instead. If you want your favorite plot on your home page, or need to document your work on WWW, then there is a switch `print -dtiff file.tif` available. The TIFF format is accepted in HTML along with the options `-dpng` and `-djpeg`. The help documentation for `print` contains more options and detailed information. See also the table on page 26.

- If you, for sake of reproducibility, want to include the MATLAB code in the document, all m-files can be inserted as regular text in the document. A further option, is to use the diary function as described in Section 4. The command `diary file` initiates the diary, saving all subsequent input and MATLAB output to the specified file. Typing `diary off` turns the function off, and the resulting text file can be imported into your document.

- Animations can be generated by recording several figure frames in sequence using the command `movie`. Exportable movies in the **AVI** format are obtained by `movie2avi`. Figure trajectories are animated by the commands `comet` and `comet3`.

Reproducibility of documentation is vital in larger projects and this is particularly important in projects involving several coworkers or projects spanning over a long time. There will always sooner or later be requirements to change graphics with respect to axis, fonts, colors and so on. The general advice is to save all data and complete associated m-files for generating all graphics. An example of structuring the files is given below:

- Create two subdirectories, one for m-files and one for figures.

- Let each example you produce or simulation you perform be represented by one script m-file. This script m-file sets up all simulation parameters and calls other m-files to perform the simulation. The simulation result is displayed in a figure plot also reproduced by this script m-file. Finally, save the figures in the figure subdirectory using the `print` command.

This general guideline will also prove convenient when similar plots are needed in another report or presentation. A short time spent on organization of the material and file administration will pay off later on. As a recommendation, save figures and its generating m-files with the same name, as outlined below.

```
>> pwd
C:\report
>> dir figures
fig1.m    fig1.eps    fig1.tif
fig2.m    fig2.eps    fig3.m
fig3.eps
```

MS Word users have the possibility to use the MATLAB **notebook** environment. This feature provides a living document where the reader can confirm the results of the MATLAB computations and make small changes in the code, directly in the document. This is a very appealing option for technical documentation and education, but perhaps the number of applications of the notebook concept is smaller than can be expected. There is a section in the Matlab pdf on-line documentation (direct link from **helpdesk**) on the use of notebooks.

A guided tour 65 (Documentation)

There are a few possibilities for making the presentation more alive. Instead of fixed plots, there is a comet plot that animates the drawing of the figure. Simply change **plot** to **comet**, or **plot3 comet3**, and the result is that the plot appears like a comet with a tail in a color different from its history. The code shows an example where a random walk in two dimensions is illustrated both as an animated comet and as a standard plot. Try the code to the right to see how **comet** works.

```
>> X=cumsum(randn(200,1));
>> Y=cumsum(randn(200,1));
>> comet(X,Y)
>> plot(X,Y)
>> title('2D random walk')
```

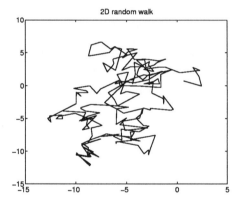

As an alternative to `comet` (which usually runs too fast), we will show how to make a similar animation using the standard plot function. The trick to get a *flimmer* free animation is to use the `EraseMode` property, and just changing the (x, y) data in the plot.

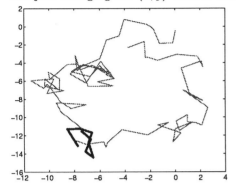

It is very easy to generate a real animation as well. The `movie` function utilizes a `struct array`, and each plot is saved as a frame in `M.cdata`. The other field `M.colormap` is default empty. Before Release 10, a matrix was used instead, and the movie had to be initialized with `M = moviein(20)`.

The sequence of plots of a bouncing ball is here animated 10 times. For multimedia presentations, exportable movies can be included in programs as Powerpoint. In recent releases of MATLAB, an *AVI* movie file can be created from a MATLAB movie with the `movie2avi` function. There are also free-ware tools for creating movies in for instance *MPEG*. Search for the function `mpgwrite` on the Mathworks' home page.

File name: `randomwalk.m`

```
N=100;  % Size of walk
n=10;   % Size of tail
x=cumsum(randn(N,1));
y=cumsum(randn(N,1));
plot(x,y,'-r');
hold on
h=plot(x,y,'-b');
set(h,'Erasemode','xor',...
      'LineWidth',4)
for i=n:N
   set(h,'XData',x(i-n+1:i),...
   'YData',y(i-n+1:i))
   pause(0.1)
end
hold off
```

File name: `bounce.m`

```
[X,Y,Z]=sphere(20);  % Unit ball
figure, axis([-3 3 -3 3 0 6])
for j=1:20;          % Bounce
   surf(X,Y,...
      Z+abs(5*sin(pi*j/20)),Z)
   M(:,j)=getframe;
end
movie(M,10)          % Animate!
```

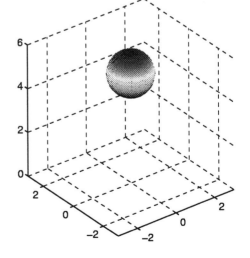

Appendix A

ANSWERS TO THE EXERCISES

Solution to Exercise 1

```
>> format long
>> pi
ans =
   3.14159265358979
>> exp(1)
ans =
   2.71828182845905
```

Solution to Exercise 2

```
>> format rat
>> pi
ans =
   355/113
>> exp(1)
ans =
   1457/536
```

Solution to Exercise 3

```
>> pi^exp(1)
ans =
   22.4592
>> exp(1)^pi
ans =
   23.1407
```

That is, $e^\pi > \pi^e$.

Solution to Exercise 4

```
>> x1=-3/2+sqrt( (3/2)^2-1)
x1 =
   -0.3820
>> x2=-3/2-sqrt( (3/2)^2-1)
x2 =
   -2.6180
>> x1^2+3*x1+1
ans =
   1.1102e-016
>> x2^2+3*x2+1
ans =
    0
```

Solution to Exercise 5

```
>> exp(i*pi)+1
ans =
        0 +1.2246e-016i
```

Solution to Exercise 6

```
>> phi=0; exp(i*phi)-cos(phi)-i*sin(phi)
ans =
    0
>> phi=pi/4; exp(i*phi)-cos(phi)-i*sin(phi)
ans =
    0
>> phi=pi/2; exp(i*phi)-cos(phi)-i*sin(phi)
ans =
    0
```

Solution to Exercise 7

```
>> i^i
ans =
    0.2079
>> exp(-pi/2)
ans =
    0.2079
```

Solution to Exercise 8

```
>> A = diag(1:10);
>> A(1,:) = 1:10;
>> A(:,1) = [1:10]';
>> M = [2*eye(2) zeros(2,3);
        ones(2,3)*3 eye(2)*4];
```

Solution to Exercise 9

```
>> m=7; n=8; A=1:n*m; A=reshape(A,m,n)'
```

Solution to Exercise 10

```
>> n=3; A=magic(n), (n^3+n)/2
A =
     8     1     6
     3     5     7
     4     9     2
ans =
    15
>> [sum(A) sum(A')]
ans =
    15    15    15    15    15    15
>> n=4; A=magic(n), (n^3+n)/2
A =
    16     2     3    13
     5    11    10     8
     9     7     6    12
     4    14    15     1
ans =
    34
>> [sum(A) sum(A')]
ans =
    34    34    34    34    34    34    34    34
```

Solution to Exercise 11

```
>> n=3; A=magic(n); v=ones(n,1);
>> s=(n^3+n)/2; (A*v-s*v)'
ans =
     0     0     0
>> n=4; A=magic(n); v=ones(n,1);
>> s=(n^3+n)/2; (A*v-s*v)'
ans =
     0     0     0     0
```

Solution to Exercise 12

```
>> n=4; A=magic(n); v=ones(n,1);
>> s=(n^3+n)/2; P=A/s;
>> P^10
ans =
    0.2500    0.2500    0.2500    0.2500
    0.2500    0.2500    0.2500    0.2500
    0.2500    0.2500    0.2500    0.2500
    0.2500    0.2500    0.2500    0.2500
```

Solution to Exercise 13

```
>> [(1:10)' cumprod(1:10)']
ans =
         1          1
         2          2
         3          6
         4         24
         5        120
         6        720
         7       5040
         8      40320
         9     362880
        10    3628800
```

Solution to Exercise 14
Since linear combinations of Gaussian random variables are Gaussian distributed and $P = R^T R$ with $R = \left(\begin{smallmatrix} 4 & 2 \\ 2 & 4 \end{smallmatrix}\right)$ we can generate the vector according to

```
>> mu = [1;2];
>> R = [4 2;2 4];
>> x = mu + R*rand(2,1)
x =
    1.3137
    2.5679
```

Check analytically that the mean and covariance are correct.

Solution to Exercise 15
First alternative

```
>> u = 1:100;
>> v = ones(100,1);
>> u*v
ans =
        5050
```

Note the dimensions, what is v*u? Second alternative

```
>> sum(1:100)
ans =
        5050
```

Solution to Exercise 16

```
>> sqrt(sum(6./(1:100000).^2))
ans =
    3.1416
```

which is close to the analytical result
$\lim_{n\to\infty} \sqrt{S_n} = \pi$.

Solution to Exercise 17

```
>> r=floor(1001*rand(100,1));
>> [rmax,imax] = max(r);
>> r(imax) == rmax
ans =
     1
>> rsort = sort(r);
>> rsort(1) == min(r)
ans =
     1
>> rsort(100) == max(r)
ans =
     1
>> i900 = find(r>900);
>> length(i900)
ans =
     6
>> sum(r)
ans =
       46456
```

Solution to Exercise 18
We construct a uniform grid over the integration range and approximate the integral with a sum

```
>> dt = 0.001;
>> t = 0:dt:1;
>> f = 1./sqrt(t.^2+2);
>> sum(f)*dt
ans =
    0.6591
```

Compare this numerical result with the anaytical solution $\log(\frac{1+\sqrt{3}}{\sqrt{2}})$. Also compare with using a trapetzodial numerical integration trapz.

Solution to Exercise 19

```
>> dx = 0.01;
>> x = 0:dx:2*pi;
>> f = abs(sin(x));
>> I = find(f > 0.9);
>> length(I)*dx/(2*pi)
ans =
    0.2881
```

Solution to Exercise 20

```
>> A = ['  one';'  two';'three']
A =
  one
  two
three
```

Solution to Exercise 21

```
>> ind=(1:255)';
>> [num2str(ind) char(ones(255,1)*' ')
   char(ind)]
```

Solution to Exercise 22

```
>> s=char(floor(26*rand(1,17000)+65));
>> i = find(s == 'A');
>> length(i)
ans =
    682
>> 17000/26
ans =
    653.8462
```

Solution to Exercise 23
Consider the short example with five TLA's

```
>> s = char(round(26*rand(5,3)+65));
>> s1 = s(:,1);
>> [s1sort,ind] = sort(s1);
>> ssort = s(ind,:)
ssort =
BCE
DGJ
ORD
PLS
UCJ
```

Solution to Exercise 24

```
>> str = 'The spy must die';
>> tmp = reshape(str,[4 4])';
>> codestr = tmp(:)'
codestr =
Tsm hpudeysi  te
>> tmp = reshape(codestr,[4 4])'
tmp =
Tsm
hpud
eysi
  te
>> decodestr = tmp(:)'
decodestr =
The spy must die
```

Solution to Exercise 25

```
>> sdecoded=char(64+mod(cumsum(...
   [double(scoded)-64]),26))
sdecoded =
THESPYMUSTDIE
```

Solution to Exercise 26

```
>> c=[];
>> N=[10; 20; 50; 100; 200; 500; 1000; 2000
   5000; 10000; 20000; 50000]';
>> for n=N
     p=primes(n); key=prod(p(end-1:end));
     tic; for i=1:100; factor(key); end
     c=[c toc];
end
>> plot(N,c-c(1))
```

The for loop avoids rounding off the system clock to zero, and the subtraction with c(1) compensates for this loop.

Solution to Exercise 27
Just subtract the random numbers instead!

```
>> rand('seed',key)
>> sveccoded=double(scoded);
>> e=round(256*rand(size(svec)));
>> svecdecoded=mod(sveccoded-e,256);
>> sdecoded=char(svecdecoded)
sdecoded =
The Spy must die
Please, confirm!
```

Solution to Exercise 28

```
>> x = -0.5:0.001:1.5;
>> f = 1./((x-0.3).^2+0.01) + ...
       1./((x-0.9).^2+0.04) -6;
>> plot(x,f)
>> disp('use the mouse cursor and');
>> disp('click on local maximum and minima');
>> [x,y]=ginput(3)
```

Solution to Exercise 29

```
>> t = 0:0.01:8*pi;
>> x = cos(-11.*t./4)+7*cos(t);
>> y = sin(-11.*t./4)+7*sin(t);
>> plot(x,y)
```

Solution to Exercise 30

```
>> t=0:0.001:10;
>> w = 1;
>> f = t.*cos(w*t)+i*t.*sin(w*t);
>> plot(f,'b');
>> hold on
>> w = 2;
>> f = t.*cos(w*t)+i*t.*sin(w*t);
>> plot(f,'r');
>> w = 3;
>> f = t.*cos(w*t)+i*t.*sin(w*t);
>> plot(f,'g');
>> hold off
>> axis square
```

Solution to Exercise 31

```
>> t=0:0.001:10;
>> wc = 10;
>> ws = 1;
>> sAM = sin(ws*t).*sin(wc*t);
>> sFM = sin(wc*t+5*sin(ws*t));
>> subplot(211)
>> plot(t,sAM)
>> title('AM')
>> subplot(212)
>> plot(t,sFM)
>> title('FM')
```

Solution to Exercise 32

```
>> x = rand(1000,2);
>> plot(x(:,1),x(:,2),'.')
>> title('Dist. of 1000 uniform samples')
>> axis square
>> figure
>> x = randn(1000,2);
>> plot(x(:,1),x(:,2),'.')
>> grid
>> title('Dist. of 1000 normal samples')
>> axis([-3 3 -3 3])
>> axis square
```

Solution to Exercise 33
Good approximation for $n > 8$.

```
>> n=2:100;
>> plot(n,cumprod(n-1)./...
   (sqrt(2*pi)*n.^(n-0.5).*exp(-n)))
>> hold on
>> plot([1 100],[1.01 1.01],'--')
>> plot([1 100],[0.99 0.99],'--')
>> hold off
```

Solution to Exercise 34

```
>> A = randn(3,3);
>>
>> B = rand(3,3);
>>
>> det(eye(3)+A*B)-det(eye(3)+B*A)
ans =
  -1.3323e-15
>>
>> trace(A*B)-trace(B*A)
ans =
   1.1102e-16
```

Solution to Exercise 35

```
>> A = magic(3);
>> B = [1 0 0]';
>> C = 2; D = B';
>> LHS = inv(A+B*C*D)
LHS =
    0.1137   -0.1116    0.0494
   -0.0472    0.0086    0.1116
   -0.0150    0.1845   -0.1009
>> RHS = inv(A)-inv(A)*B*...
   inv(inv(C)+D*inv(A)*B)*D*inv(A)
RHS =
    0.1137   -0.1116    0.0494
   -0.0472    0.0086    0.1116
   -0.0150    0.1845   -0.1009
```

Solution to Exercise 36

```
>> A = magic(3);
>> B = [1 0 0]';
>> C = 2; D = B';
>> LHS = inv([A B;D C]);
>> Ai = inv(A);
>> Deli = inv(C-D*Ai*B);
>> RHS = [Ai+Ai*B*Deli*D*Ai -Ai*B*Deli;
          -Deli*D*Ai Deli];
>> LHS-RHS
ans =
  1.0e-16 *
    0.2776        0        0        0
         0        0        0        0
   -0.0694        0   0.1388  -0.0173
         0        0        0        0
```

Solution to Exercise 37

```
>> r = [1;2]; p = poly(r);
>> r == roots(poly(r)),
ans =
     0
     0
>> p == poly(roots(p))
ans =
     1     1     1
```

Solution to Exercise 38

```
>> r = 1:6;
>> p = poly(r);
>> x = 0:0.01:7;
>> plot(x,polyval(p,x))
>> % Analytic differentiation
>> pprim = p(1:6).*[6:-1:1]
pprim =
   6   -105   700   -2205   3248   -1764
>> roots(pprim)
ans =
   5.6634
   4.5737
   3.5000
   2.4263
   1.3366
```

Solution to Exercise 39

```
>> d = 0.1;
>> [S,T] = meshgrid(-pi/2:d:pi/2,-pi/2:d:pi/2);
>> X = tan(S);
>> Y = tan(T);
>> f = atan(3.*X.*exp(Y)-X.^3-3.*exp(3.*Y));
>> mesh(S,T,f);
>> rotate3d on
```

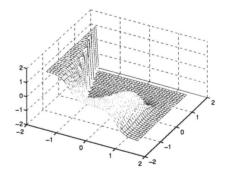

Solution to Exercise 40

```
>> [X,Y]=meshgrid(-1-eps:0.1:1,-1-eps:0.1:1);
>> f = X.*Y.*(X.^2-Y.^2)./(X.^2+Y.^2);
>> f(10,10) = 0;
>> mesh(X,Y,f);
```

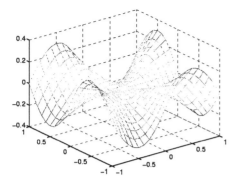

Solution to Exercise 42

The file can contain the following two lines:

```
s=pwd;
addpath(s);
```

Solution to Exercise 45

File name: hailstorm.m

```
function x=hailstorm(x0)
% Compute the hailstorm series for x(1)=x0
x(1)=x0;
n=1;
while x(n)~=1
  if x(n)/2==round(x(n)/2)
      x(n+1)=x(n)/2;
  else
      x(n+1)=3*x(n)+1;
  end
  n=n+1;
end
```

Solution to Exercise 46

File name: ratpi.m

```
function [p,q]=ratpi(n)
err=1;
for pp=1:10^n
  for qq=1:10^n
    if abs(pi-pp/qq)<err
        err=abs(pi-pp/qq);
        p=pp; q=qq;
    end
  end
end
```

Solution to Exercise 47

File name: ratapprox.m

```
function [p,q]=ratapprox(x,tol)
err=1;
q=1;
while err>tol
    q=q+1;
    p=round(x*q);
    err=abs((x-p/q)/x);
end
```

Solution to Exercise 48

File name: fibonacci.m

```
function y=fibonacci(n,method)
% Compute Fibonacci number n
switch method
  case 1, y=fib1(n);
  case 2, y=fib2(n);
  case 3, y=fib3(n);
end

function y=fib1(n)
y=[0 1];
for i=3:n
  y(i)=y(i-1)+y(i-2);
end
y=y(n);

function y=fib2(n)
if n==1
    y=0;
elseif n==2
    y=1;
elseif n>2
    y=fib2(n-1)+fib2(n-2);
end

function y=fib3(n)
x=[0;1];
A=[1 1;1 0];
xn=A^(n-1)*x;
y=xn(1);
```

Compare the computation times!

Solution to Exercise 49

File name: mysqrt.m

```
function y=mysqrt(x,y0,tol)
y=y0; y0=Inf;
while abs(y-y0)>tol
    y0=y;
    y=(y0+x/y0)/2;
end
```

Solution to Exercise 50

File name: `mcnugget.m`

```
function N=mcnugget(n)
% non-McNugget numbers < n
N=1:n;
for i=0:n/6
   for j=0:n/9
      for k=0:n/20
         x=i*6+j*9+k*20;
         ind=find(N==x);
         if ~isempty(ind);
            N(ind)=[];
         end
      end
   end
end
```

The answer is 43!

Solution to Exercise 51

File name: `mandelbrot.m`

```
maxiter=20;
c1=-2:0.05:1;
c2=-1.5:0.05:1.5;
for m=1:length(c1)
   for n=1:length(c2)
      c=c1(m)+i*c2(n);
      N(m,n)=0; z=0;
      while abs(z)<2 & N(m,n)<maxiter
         N(m,n)=N(m,n)+1;
         z=z^2+c;
      end
   end
end
pcolor(c1,c2,N)
```

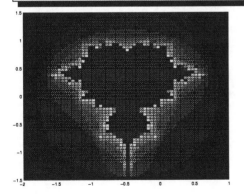

Solution to Exercise 52

File name: `fdiff2.m`

```
function fprime=fdiff2(f,x,h)
%fprime=fdiff(f,x)
if nargin < 3
   h = 0.01;
end
fprime = eval(['(',f,'(x+h) - ',f,'(x))/h']);
```

Solution to Exercise 53

```
>> f = inline('1/((x-0.3)^2+.01)+1/((x-0.9)^2+0.04)-6');
>> fv = vectorize(f);
>> x = -1:0.01:2;
>> plot(x,fv(x));
>> grid
>> finv = inline('-1/((x-0.3)^2+.01)-1/((x-0.9)^2+0.04)+6');
>> xmax = fminbnd(finv,0,0.5)
xmax =
    0.3004
>> xz1 = fzero(f,0)
xz1 =
   -0.1316
>> xz2 = fzero(f,1)
xz2 =
    1.2995
```

Solution to Exercise 54

File name: `myint.m`

```
function S = myint(f,a,b,N)
% S=int(f,a,b,N)

T=(b-a)/N;
x=a + [0:T:(N-1)*T];
S = sum(feval(f,x)*T);
```

```
>> myint(inline('x.*exp(x.^2)'),0,3,10)
ans =
    1.4880e+03
>> myint(inline('x.*exp(x.^2)'),0,3,1e4)
ans =
    4.0474e+03
>> myint(inline('x.*exp(x.^2)'),0,3,1e6)
ans =
    4.0510e+03
>> quad(inline('x.*exp(x.^2)'),0,3)
ans =
    4.0510e+03
```

That is, for this particular function, where most of
the area is located close to 3, it requires quite a
large number of intervals. The MATLAB function
quad uses an adaptive interval length and is faster.

Solution to Exercise 55
File name: `findzero.m`

```
function x = findzero(f,x0,tol,h)
%x = findzero(f,x0,tol,h)

x=x0+2*tol;
while(abs(x-x0)>tol)
  x0 = x;
  x = x - feval(f,x)/...
     ((feval(f,x+h)-feval(f,x))/h);
end
```

Solution to Exercise 56
File name: `fdiff3.m`

```
function fprime=fdiff3(f,x,ep)
%fprime==fdiff3(f,x,ep)
h = 0.1;
f1 = fdiff(f,x,h);
h = h * 0.99;
f2 = fdiff(f,x,h);
while(abs(f1-f2)>ep*f2)
  h = h*0.99;
  f1 = f2;
  f2 = fdiff(f,x,h);
end
fprime = f2;

function fd = fdiff(f,x,h)
fd = (feval(f,x+h)-feval(f,x))/h;
```

Solution to Exercise 59

```
>> M = [1 1;1 2];
>> v = [2;3];
>> x = M\v
x =
     1
     1
>> a = 0:0.1:5;
>> b = x(1)+x(2)*a;
>> plot(a,b); xlabel('a'); ylabel('b')
>> hold on, plot(1,2,'b*',2,3,'b*')
>> M = [M; 1 4]; v = [v;6];
>> x = M\v
x =
    0.5000
    1.3571
>> b = x(1)+x(2)*a;
>> plot(a,b,'--'), plot(4,6,'*')
```

```
>> title('Least squares line fit')
```

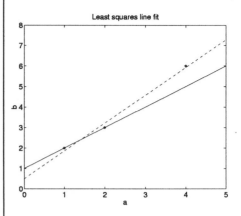

Solution to Exercise 73
$$[L, U] = [-1023, 1024]$$

Solution to Exercise 74

```
>> 2^(1023)*(2-eps)   % realmax
>> 2^(1-1023)         % realmin
```

Solution to Exercise 75

```
>> sin(pi)
ans =
   1.2246e-016
>> 0-exp(987)
ans =
  -Inf
>> exp(986)-exp(987)
ans =
   NaN
>> 1+1e-17
ans =
     1
```

The results depend on
1) pi is stored with limited precision.
2) e^{987} is larger than `realmax`, that is, infinity Inf.
3) Inf-Inf is defined as Not-a-Number NaN.
4) 10^{-17} is less than machine precision eps, so it is rounded off to zero.

Appendix B

COMMAND REFERENCE

MATLAB features thousands of functions divided in more than 20 different categories. In this appendix we list a subset of these categories and a selection (basically the ones described in this book) of the functions in each category. Typing `help` followed by the category name gives a complete list of the functions in each category. Likewise, typing `help` followed by the function name gives detailed information regarding a specific function. The same information can also be reached from the hyperlinked help information available threw the command `helpdesk`.

MATLAB'S MAIN CATEGORIES OF FUNCTIONS	
general	General purpose commands
ops	Operators and special characters
lang	Language constructs and debugging
elmat	Elementary matrices and matrix manipulation
elfun	Elementary math functions
specfun	Specialized math functions
matfun	Matrix functions – numerical linear algebra
datafun	Data analysis and Fourier transforms
polyfun	Polynomial and interpolation functions
funfun	Function functions – Nonlinear numerical methods
sparfun	Sparse matrix functions
datatypes	Data types and structures
strfun	Character string functions
iofun	Low-level file I/O functions
timefun	time and date functions
graph2d	Two dimensional graphics
graph3d	Three dimensional graphics
timefun	Time and date functions
specgraph	Specialized graphical functions
uitools	Graphical user interface creation

Polynomials & Interpolation – polyfun

POLYNOMIALS	
poly	polynomial with specified roots ...31
polyval	evaluate polynomial 28
roots	polynomials roots 31
residue	partial fraction expansion

INTERPOLATION FUNCTIONS	
interp1	one-dimensional data interpolation
interp2	two-dimensional data interpolation
meshgrid	grid data for plots 33
spline	cubic spline interpolation

General purpose commands – general

MANAGING FUNCTIONS	
help	on-line help information 2
helpdesk	show help in web browser 2
lookfor	search for keyword 43
type	display function definition 4, 18
edit	start editor/debugger 40
which	function search path 4, 39
what	list files in directory
ver	display installed versions 2
path	alter the search path 39
pathtool	interactive path browser 39
addpath	add directories to search path ...39
rmpath	remove directories from path 39

MANAGING THE WORKSPACE	
who, whos	list defined variables 17, 18
workspace	workspace browser 18
save	save variables to disk 18
load	load variables from disk 18
clear	clear variables and functions from memory 18
size	array size 11
length	maximal size dimension 11
disp	display array 16
format	command window format 6

OPERATING SYSTEM INTERACTION	
cd	change working directory 38
pwd	print working directory 39
diary	save session to text file 18
dir	directory listing 38
mkdir	make directory 38
matlabroot	MATLAB root directory
!	operating system commands ..38
quit	quit MATLAB 2
startup	startup m-file 40

Operators and special characters – ops

ARITHMETIC OPERATORS – arith	
+, -	array addition and subtraction
*	matrix multiplication
.*	array multiplication 11
^	matrix power 11
.^	array power 11
kron	Kronecker tensor product

MATRIX DIVISION – slash	
\	backslash or left division 30
/	slash or right division
./	array division 11

PUNCTUATION – punct	
:	index to arrays or regularly spaced vector, see online help for colon 9, 12
...	expression continuation 51
'	conjugate transpose or quote text . 10, 16
.'	non-conjugated transpose 10
%	comment 7

RELATIONAL OPERATORS – relop	
==	equality 12
< >	relational operators 12
&	logical and 12
		logical or 12
~	logical not 12
xor	logical exclusive or 12

LOGICAL FUNCTIONS	
find	find indices of nonzero values 12
all	test if all elements are nonzero 12
any	test if any non-zeros 12

Specialized Math Functions – specfun

SPECIALIZED MATH FUNCTIONS – specfun	
factorial	factorial function
pow2	base 2 power
rat, rats	fraction approximation

Language Constructs – lang

CONTROL FLOW	
if, elseif	conditional execution 46
else	conditional execution 46
end	terminate control flow statements or indicate last array index ... 46
for	repetition statement 48
while	conditional repetition 52
switch	switch among cases 46
case	case switch 47
otherwise	default part of case switch 47
break	terminate loop execution 52
return	return to invoking function ... 52
error	display error message 46

INTERACTIVE INPUT AND DEBUGGING	
input	request user input 46
keyboard	invoke keyboard in m-file 54
menu	generate user menu
pause	halt execution temporarily
dbstop	set breakpoints 55
dbclear	clear breakpoints 55

Data Types and Structures – datatypes

STRUCTURES AND CELL ARRAYS – datatypes	
struct	create structure array
fieldnames	field names of structure array
rmfield	remove structure fields
setfield	set contents of field
getfield	get contents of field
struct2cell	convert structure to cell array
cell	create cell array
num2cell	convert numeric array to cell array

MULTIDIMENSIONAL ARRAYS – datatypes	
cat	concatenate arrays
flipdim	flip array along dimension
permute	rearrange dimensions
reshape	reshape array
squeeze	remove singleton dimensions

Elementary Matrices – elmat

ELEMENTARY MATRICES	
blkdiag	block diagonal matrix
eye	identity matrix
ones	all ones array
zeros	all zeros array
rand	uniformly distributed random array
randn	normally distributed random array
linspace	linearly spaced vectors
logspace	logarithmically spaced vectors

Time and Date Functions – timefun

TIMES AND DATES	
clock	current time
cputime	elapsed CPU time 50
date	current date string
etime	elapsed time 50
tic, toc	stopwatch timer 50

CONSTANTS	
ans	answer from previous expression ... 6
eps	floating point precision 6
realmax	largest floating point number 6
realmin	smallest floating point number 6
flops	count floating point operations (obsolete)
pi	π 7
i, j	imaginary unit $\sqrt{-1}$ 8
Inf	infinity 6
NaN	Not-a-Number 7, 123

MATRIX MANIPULATION	
cat	concatenate arrays
diag	diagonal matrices and diagonals of a matrix
fliplr	mirror matrix vertically
flipud	mirror matrix horizontally
repmat	replicate matrix
reshape	reshape array
rot90	rotate matrix 90 degrees
tril	lower triangular part of matrix
triu	upper triangular part of matrix

Elementary Functions – elfun

ELEMENTARY MATH FUNCTIONS	
abs	absolute value or complex magnitude 8
angle	phase angle 8
real, imag	real and imaginary part 8
conj	complex conjugate 8
acos, cos	cosine functions 7
asin, sin	sine functions 7
atan, tan	tangent functions 7
atan2	four quadrant inverse tangent
sqrt	square root 7
exp, log	exponential and natural logarithm 7
log2, log10	base two and base ten logarithm 7
sign	signum function
gcd	greatest common divisor
rem	remainder after integer division 49
mod	modulus, signed remainder
fix	round towards zero 6
floor	round towards $-\infty$ 6
ceil	round towards $+\infty$ 6
round	round to nearest integer 6

Function Functions – funfun

FUNCTION FUNCTIONS – funfun	
fzero	single variable nonlinear equation solver 57
fminbnd	minimize function of one variable 57
fminsearch	minimize function of several variables
quad, quad8	numerical integration 57
dblquad	numerical double integration
ode45	solve differential equations .. 57, 57
feval	evaluate a matlab function ...57
eval	evaluate a matlab expression 57

Matrix Functions – matfun

MATRIX ANALYSIS	
cond	condition number 30
rcond	reciprocal condition number
det	determinant 29
trace	sum of diagonal elements
norm	vector and matrix norm 30
null	null space
orth	range space
rank	matrix rank 29
rref	reduced row echelon form
rrefmovie	rref movie
subspace	angle between two subspaces

LINEAR EQUATIONS	
chol	Choleskey factorization
inv	matrix inverse 29
lu	LU factorization of matrix
lsqnonneg	nonnegative least squares
pinv	pseudo inverse
qr	orthogonal-triangular factorization

EIGENVALUES AND MATRIX FUNCTIONS	
eig	eigenvalues and eigenvectors 29
svd	singular value decomposition 29
schur	Schur decomposition
expm	matrix exponential 31
logm	matrix logarithm
sqrtm	matrix square root 31
funm	evaluate functions of a matrix

Character String Functions – strings

CHARACTER STRING FUNCTIONS – strings	
char	create character array 16
double	convert character array to number 16
str2double	convert string to double
str2num	convert string to number 16
num2str	convert number to string
strcmp	compare strings 17
upper	upper case conversion 17
lower	lower case conversion 17

Data Analysis Functions – datafun

BASIC OPERATIONS	
factor	prime factors
primes	list prime numbers
max, min	maximum and minimum entries .13
sum	sum of array elements 13
prod	product of array elements 13
trapz	trapezoidal numerical integration
mean	mean value of array
median	median value of array
std	standard deviation
convhull	convex hull
cumprod	cumulative product 13
cumsum	cumulative sum 13
cumtrapz	cumulative trapz
sort	sort elements in ascending order .13
sortrows	sort rows in ascending order
diff	differences 13
del2	discrete Laplacian
gradient	numerical gradient

FILTERING AND TRANSFORMS	
conv	convolution and polynomial multiplication
deconv	deconvolution and polynomial division
conv2	two-dimensional convolution
filter	linearly filter data
filter2	two-dimensional filtering
fft	fast Fourier transform 153
ifft	inverse FFT
fft2	two-dimensional FFT
ifft2	inverse two-dimensional FFT
fftshift	shift frequency axis

SOUND PROCESSING FUNCTIONS – datafun	
sound	convert vector into sound
soundsc	scale data and play sound
auread	read .au file
auwrite	write .au file
wavread	read .wav file
wavwrite	write .wav file
wavrecord	record .wav file on PC
wavplay	play .wav file on PC

Data Visualization

BASIC PLOT FUNCTIONS – **graphics**	
`plot`	plot vectors and matrices 22
`bar, barh`	vertical and horizontal bar chart 37
`pie`	pie chart plot 37
`hist`	plot histograms 53
`polar`	polar coordinate plot
`semilogx`	plot in semi log scale 22
`semilogy`	plot in semi log scale 22
`loglog`	plot in log-log scale 22
`hold`	hold current plot 24
`print`	send current figure to the printer 26
`clf`	clear current figure 23
`subplot`	create axes in tiled positions ... 26

SPECIALIZED PLOTTING – **specgraph**	
`comet`	comet plot 167
`pcolor`	pseudo-color checkerboard plot ... 131
`quiver`	velocity plot 145
`stem`	discrete sequence plot 23

THREE-DIMENSIONAL PLOTTING – **graph3d**	
`bar3, bar3h`	bar plots
`comet3`	comet plot 167
`plot3`	three-dimensional plots
`stem3`	plot discrete surface data
`quiver3`	three-dimensional velocity plot
`waterfall`	waterfall plot

SURFACE PLOTS – **graph3d**	
`contour`	plot level curves 33
`contourc`	contour computation
`mesh`	mesh plot 33
`meshc`	combination mesh and contour ..33
`surf`	shaded surf plot 33
`surfl`	surface plot with lighting 33
`surfc`	combination surf and contour ...33

PLOT ANNOTATION – **graph2d, graph3d**	
`clabel`	contour labels 34
`colorbar`	display color scale 33
`grid`	grid lines in 2D and 3D plots 25
`gtext`	place text in 2D plots using the mouse 25
`legend`	graph legend for lines and patches 25
`title`	add title text to 2D and 3D plots 25
`axis`	change the axis scaling 25
`xlabel`	label for x-axis in 2D and 3D plots 25
`ylabel`	label for y-axis in 2D and 3D plots 25
`zlabel`	label for z-axis in 3D plots

Appendix C

SUMMARY OF MATHEMATICAL FUNCTIONS

Matrix Algebra Elementary Functions			
Component-wise computation	Component-wise definition	Matrix computation	Matrix definition
`A=B.*C`	$(A)_{i,j} = (B)_{i,j} \cdot (C)_{i,j}$	`A=B*C`	$A = B \cdot C$
`A=B./C`	$(A)_{i,j} = (B)_{i,j}/(C)_{i,j}$	`A=B/C`	Implicit by $B = C \cdot A$
`A=B.^k`	$(A)_{i,j} = (B)_{i,j}^{k}$	`A=B^k`	$A = B^k = B \cdot B \cdots B$
`A=sqrt(B)`	$(A)_{i,j} = \sqrt{(B)_{i,j}}$	`A=sqrtm(B)`	Implicit by $B = A^2$
`A=exp(B)`	$(A)_{i,j} = e^{(B)_{i,j}}$	`A=expm(B)`	$A = e^B, \quad \frac{d}{dt}e^{Bt} = Be^{Bt},$ $e^B = e^{Bt}\big|_{t=1} = \sum_{k=0}^{\infty} \frac{1}{k!}B^k$
`A=log(B)`	$(A)_{i,j} = \log(B)_{i,j}$	`A=logm(B)`	Implicit by $B = e^A$

Table 27.1 Elementary operations with dual meaning for matrix algebra.

Matrix Algebra	
MATLAB	Definition
lambda=eig(A) [V,D]=eig(A)	The *eigenvalue* λ_i with *eigenvector* v_i. They are defined as $A\lambda_i = \lambda_i v_i$ for $i = 1, 2, \ldots n$. $\lambda = (\lambda_1, \lambda_2, \ldots, \lambda_n)$ are sorted in descending order and $V = [v_1, v_2, \ldots, v_n]$, $D = \text{diag}(\lambda)$. Spectral theorem for symmetric matrices: $A = \sum_i \lambda_i v_i v_i^T$.
sigma=svd(A) [U,S,V]=svd(A)	The *singular value decomposition*. It is defined as $A = U\Sigma V^T$, where U, V are unitary, so $U^T U = I$ and $\Sigma = U^T A V$. $\sigma = \text{diag}(\Sigma)$ are the *singular values* sorted in descending order $\sigma_1 \geq \sigma_2 \geq \cdots \geq \sigma_n \geq 0$. The *spectral theorem* for arbitrary matrices: $A = \sum_i \sigma_i u_i v_i^T$. Relation to eigenvalues: We have for the symmetric matrix $A^T A = V\Sigma^2 V^T$, so $\sigma_i^2(A) = |\lambda_i(A)|^2$.
[Q,R]=qr(A)	The *QR decomposition* computes an upper triangular matrix R of the same size as A, and a unitary matrix Q (that is, a square matrix with the properties $Q^T Q = I$ and $QQ^T = I$) such that $A = QR$.
N=null(A)	Computes an orthonormal base for the *null space* of the matrix A, such that $N^T N = I$ and $N^T A = 0$.
p=poly(A)	Computes the characteristic polynomial of the matrix A, defined by the polynomial $p(\lambda) = \det(\lambda I - A)$. We have that poly(A)=poly(eig(A)).
norm(A)	The *matrix norm* is defined as the largest singular value σ_1, s=svd(A); s(1).
cond(A)	The *condition number* is defined as the ratio of the largest and smallest singular values σ_1/σ_n, s=svd(A); s(1)/s(end).
rank(A)	The *rank* of a matrix is defined as the number of linearly independent rows or columns, and can be computed as the number of non-zero singular values s=svd(A); sum(s>0).

Table 27.2 Basic definitions in matrix algebra.

Optimization		
MATLAB	Toolbox	Definition
`thhat=lsqcurvefit(f,th0,X,Y)`	`optim`	The *least squares curve fit* computes the parameter $\hat{\theta}$ that gives the best fit to data of the parametric function $y = f(x; \theta)$, given the data vectors X and Y. The algorithm is initialized at `th0`. Formally, $\hat{\theta} = \arg\min_\theta \sum_i |Y_i - f(X_i, \theta)|^2$
`thhat=lsqnonlin(f,th0)`	`optim`	The least squares solution to a non-linear system of equations computes the parameter $\hat{\theta}$ that minimizes the vector valued function $f(\theta)$. The algorithm is initialized at `th0`. Formally, $\hat{\theta} = \arg\min_\theta f^T(\theta)f(\theta)$
`xhat=linprog(f,A,b)`	`optim`	Solves the *linear programming* problem $\min_x f^T x$, where $Ax \le b$.
`xhat=quadprog(H,f,A,b)`	`optim`	Solves the *quadratic programming* problem $\min_x x^T H x + f^T x$, where $Ax \le b$.

Table 27.3 Basic optimization functions in MATLAB and the optimization toolbox

SAMPLING AND INTERPOLATION IN 1D		
MATLAB	Toolbox	Definition
`y=decimate(x,M);`	`signal`	Down-sampling a factor M with an anti-alias filter. Assumes regular spaced samples in `x`
`y=interp(x,M);`	`signal`	Up-sampling a factor M. Assumes regular spaced samples in `x`
`y=resample(x,P,Q);`	`signal`	Re-sampling a factor P/Q. Assumes regular spaced samples in `x`
`polyfit`	`matlab`	
`spline`	`matlab`	
`y=interp1(X,Y,x,'linear');`	`matlab`	Interpolates the value of $y(x)$ based on the data X, Y. Default is linear interpolation, and alternatives include `'cubic'` and `'nearest'`. Irregular spaced samples in `X, Y` are allowed.

Table 27.4 Interpolation in 1D.

SAMPLING AND INTERPOLATION IN 2D		
MATLAB	Toolbox	Definition
z=interp2(X,Y,Z,x,y);	matlab	As interp1, but for two-dimensional functions $z(x, y)$. Extensions to higher dimensions are found in interp3 and interpn. Usually used to produce intermediate values from a uniform, or at least monotonic, grid (X, Y).
[Xgrid,Ygrid]=meshgrid(X,Y)	matlab	Converting the vectors (X, Y) to matrices. Useful for computing the matrix $Z(X, Y)$ with vectorized notation, instead of looping through all combinations. Example: Z=1/(X.^2+Ygrid.^2); is faster than for i=1:N; for j=1:N; Z(i,j)=1/(X(i)^2+Y(i)^2); end; end
z=griddata(X,Y,Z,x,y);	matlab	Similar to interp2, but uses delaunay and does not require monotonic data in (X, Y). Usually used to generate z on a uniform grid (x, y) from a non-uniform one (X, Y), for example for using mesh.
t=delaunay(x,y);	matlab	*Delaunay interpolation* of irregularly spaced data x,y. Returns a matrix of size length(x),3 with row indices to x,y, such that the rows of t build non-overlapping triangles.
trimesh(t,x,y)	matlab	Produces a plot of the Delaunay triangulation.
tsearch(x,y,t,x1,y1)	matlab	Finds the triangle containing the point x1, y1. Based on the Delaunay triangulation.
dsearch(x,y,t,x1,y1)	matlab	Finds the closest point containing the point x1, y1. Based on the Delaunay triangulation.
ind=convhull(x,y)	matlab	Computes the smallest convex set containing all data points. Based on the Delaunay triangulation. Use plot(x(ind),y(ind)) to plot the convex hull.
voronoi(x,y,t)	matlab	Computes and plots the *Voronoi diagram*, which consists of a mesh of straight lines, where all points on each line has exactly the same distance to at least two data points.

Table 27.5 Interpolation in 2D.

CALCULUS PART 1	
FUNCTIONS	
f	The function f is either
	1. a function name (f='sin'),
	2. a string (f='(x+1)/(x-1)'),
	3. an *inline object* (f=inline('(x+1)/(x-1)')),
	4. or a *function handle* (f=@sin),
	Alternative 4 is the fastest one, followed by 1.
	Alternatives 2 and 3 are equivalent, since an inline
	object is created internally in 2.
feval	Function evaluation. f=@sin, feval(f,pi/2) will
	return 1 as answer.
INTEGRATION	
I=quad(f,x1,x2);	Computes the integral $I = \int_{x_1}^{x_2} f(x)dx$. Uses adaptive *Simpson quadrature*.
I=quadl(f,x1,x2);	As quad. Uses adaptive *Lobatto quadrature*.
I=dblquad(f,x1,x2,y1,y2);	Computes the double integral $I = \int_{x_1}^{x_2} \int_{y_1}^{y_2} f(x,y)dxdy$.
DIFFERENTIATION	
yd=diff(y)	Computes the difference of the vector y as an approximation of the derivative. The calls y=feval(f,x1:h:x2); yd=diff(y)/h; should give a good approximation of dy/dx in $[x_1, x_2]$ for small h at well-behaved points of the function $y = f(x)$.
ORDINARY DIFFERENTIAL EQUATIONS	
[t,x]=ode45(f,[0 T],x0)	Computes a numerical solution to the *ordinary differential equation* $\dot{x}_t = f(x_t)$, with implicit solution $x_t = \int_0^T f(x_s)ds$, in the points [t,x]. There are many alternatives to ode45.
POLYNOMIALS	
pd=polyder(p)	Differentiation of polynomials using the rule $p(x) = \sum_{k=0}^n p_k x^k \Rightarrow dp(x)/dx = \sum_{k=1}^n p_k k x^{k-1}$. Note that the output polynomial is one element shorter.
pi=polyint(p,c)	Integration of polynomials using the rule $p(x) = \sum_{k=0}^n p_k x^k \Rightarrow \int p(x)dx = c + \sum_{k=0}^n p_k \frac{1}{k+1} x^{k+1}$.
p=conv(p1,p2)	The *convolution* of two vectors is defined as $p(k) = \sum_m p_1(m)p_2(k-m)$, which in the case of vectors of polynomial coefficients corresponds to multiplication $p(x) = p_1(x)p_2(x)$.
[q,r]=deconv(b,a)	The *deconvolution* gives as result two vectors (polynomials) such that b=conv(q,a)+r. For polynomials, this function computes the *polynomial division*, where the result is $b(x) = q(x)a(x) + r(x)$. This is useful to convert rational functions to one polynomial and one strictly proper rational function $\frac{b(x)}{a(x)} = q(x) + \frac{r(x)}{a(x)}$.

Table 27.6 Calculus, part 1.

CALCULUS PART 2	
RATIONAL POLYNOMIALS	
`[r,p,k]=residue(b,a)`	The *partial fraction decomposition* computes $\frac{b(x)}{a(x)} = k + \sum_i \frac{r_i}{x-p_i}$ of rational functions.
`[b,a]=residue(r,p,k)`	The rational function can be recovered by $\frac{b(x)}{a(x)} = k + \sum_i \frac{r_i}{x-p_i}$ from partial fraction decomposition. Note that this is one of few examples in MATLAB of a function also implementing its inverse, depending of the number of inputs and outputs in the call.
`[q,r]=deconv(b,a)`	The *deconvolution* function **deconv** converts rational functions to one polynomial and one strictly proper rational function $\frac{b(x)}{a(x)} = q(x) + \frac{r(x)}{a(x)}$.
`[bd,ad]=polyder(b,a)`	Computes the symbolic derivative $\frac{d}{dx}\frac{b(x)}{a(x)}$.

Table 27.7 Calculus, part 2.

SYMBOLIC CALCULUS	
x=sym('x')	Symbolic x. The functions are defined as usual. Most elementary functions are overloaded, like +, *, sin, real. Example: f='arccos(x)' Using a second optional argument 'real' assures that x is treated as a real number. Alternatives include 'positive', 'unreal'.
Asym=sym(A)	Represents the numeric matrix, or scalar, A as symbolic values. For example, 1/10 is impossible to represent exactly in binary arithmetic. pi=sym('pi'), sin(pi) is another example giving exactly zero as answer. An optional second argument is one of 'r','f','e','d', see the help text.
syms x pi	Symbolic x and π as an alternative to using sym with the 'real' option.
syms	Lists all symbolic objects in workspace.
fd=diff(f,'x')	Computes the symbolic derivative $\frac{d}{dx}f(x)$. diff is overloaded the vector difference for symbolic functions.
fi=int(f,'x')	Computes the symbolic integral $\int f(x)dx$. fi=int(f,'x',a,b) computes the integral $\int_a^b f(x)dx$.
ft=taylor(f,'x')	Computes the *Taylor expansion* $f(x) = f(x_0) + \frac{1}{1!}f^{(1)}(x_0) + \frac{1}{2!}f^{(2)}(x_0) + \ldots$. Example: syms x, taylor(exp(-x),4) returns 1-x+1/2*x^2-1/6*x^3.
limit(f,x,a)	Computes the *limit* $\lim_{x \to a} f(x)$. Example: limit('sin(x)/x',x,0)=1.
symsum(f)	Computes the sum of the terms defined by f. Example: syms k , symsum(1/k^2,1,Inf) gives the result 1/6*pi^2
simple(f)	Applies a lot of tricks to simplify the expression in f. The function uses several low-level primitives as simplify, radsimp, combine, factor, expand, convert and collect.
eval(f)	Evaluates the function. Example: syms x, f='sin(x)'; x=pi/2; eval(f) will return 1.
pretty(f)	Makes a try to format the printout in a more readable way. Be sure to use a font with fixed size, like Courier.

Table 27.8 Symbolic calculus (see help symbolic).

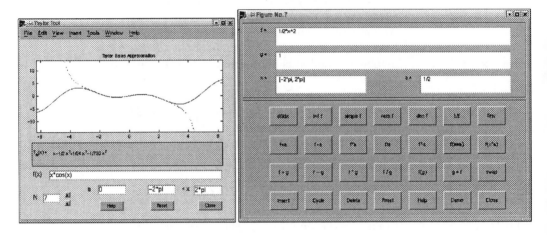

Table 27.9   GUI's for the symbolic toolbox: `taylortool` and `funtool`.

Appendix D

TOOLBOX SUMMARIES

This appendix lists the main functionalities of the most well established toolboxes. At least a couple of them may be used in under-graduate courses in the electrical engineering area. Thus, in contrast to the preceding appendix, the functions are organized in toolboxes, and some problems may have solutions in more than one toolbox. The toolboxes also have in common that they support the *LTI object*, see Section 22, for representing time-invariant linear systems:

- LTI systems may be identified from measured data, which can be done in the *System Identification Toolbox* (`help ident`).

- LTI systems are the result of filter design, as can be done in the *Signal Processing Toolbox* (`help signal`)

- LTI systems are the basis for model-based control, as covered in the *Control System Toolbox* (`help control`)

- *Simulink* (`help simulink`) is a graphical oriented tool to implement and simulate dynamical systems. Simulink does not support LTI objects directly. However, a linearized LTI model can be obtained from a block diagram by the `linmod` function, which can then be converted to an LTI object. Conversely, an LTI object is easily imported as a continuous or discrete time state space model. In this way, control design and system identification can be performed in the toolboxes, and the result imported in the simulation environment.

SIGNAL PROCESSING TOOLBOX	
GUI	
`sptool`	Multi-purpose signal processing GUI
`fdatool`	Filter design and data analysis GUI
WINDOWS	
`w=hanning(N);`	The Hanning window is used to smooth the edges of a signal. Alternative windows include `hamming,` `bartlett, blackman` or the trivial `boxcar`
COVARIANCE FUNCTION	
`Rx=xcov(x,kmax);`	Estimates the covariance function $R_x[k] = \mathrm{E}[x[n]x[n-k]]$, using $\hat{R}_x[k] = \frac{1}{N}\sum_n x[n]x[n-k]$, $k = -k_{max}, \ldots, -1, 0, 1, 2, \ldots, k_{max}$. See also `covf` in `ident`
`plot(-kmax:kmax,Rx);`	Plots the estimate
SPECTRUM ESTIMATION	
`psd(y,[],fs,gamma);`	Computes and plots Welch estimate of the spectrum, defined as $\Phi(\omega) = DFT(R_x[k])$, with sampling frequency `fs` and time window width `gamma`. The estimate is basically a smoothed version of `abs(fft(x)).^2`. See also `etfe, spa` in `ident`
`specgram(y);`	Surf plot of time varying spectrum estimate using FFT over time windowed data
FILTER DESIGN	
`fn=w/(pi/T);`	Normalized frequency $f_n \in [0,1]$, where ω is in rad/s
`fn=f/(2T);`	Normalized frequency $f_n \in [0,1]$ where f is in Hertz
`[b,a]=butter(n,fn,'low');`	Approximation of degree n of an ideal *low-pass filter* (`'low'` is default). A *high-pass filter* is obtained with `'high'`. If `fn` is a two-element vector, a *band-pass filter* is obtained with `'bandpass'` (default) and a *band-stop filter* with `'stop'` Alternatives to `butter` include `ellip, cheby1` and `cheby2`, which are syntax equivalent to butter except for that a ripple has to be specified.
`[b,a]=yulewalk(p,f,H);`	Filter of degree p approximating $B(e^{i2\pi f})/A(e^{i2\pi f}) \approx H(e^{i2\pi f})$.
`[H,f]=freqz(b,a,N,fs);`	Compute the transfer function $\frac{B(e^{i2\pi f})}{A(e^{i2\pi f})}$ in N frequency points in $[0, f_s/2]$.
`plot(f,abs(H))`	Plots the amplitude curve of the filter $|H(e^{i2\pi f})|$. Also try `semilogy, semilogx, loglog`, where the latter gives a Bode diagram.

Table 27.1 Signal processing toolbox (`help signal`)

CONTROL SYSTEM TOOLBOX AND THE LTI OBJECT, PART I	
GUI	
`ltiview`	GUI for LTI objects
SYSTEM DEFINITION AS LTI OBJECT	
`g=ss(A,B,C,D,T);`	Discrete time state space model with sampling interval T, $x[k+1] = Ax[k] + Bu[k], \quad y[k] = Cx[k] + Du[k]$.
`g=ss(A,B,C,D);`	Continuous time state space model, $\dot{x}(t) = Ax(t) + Bu(t), \quad y(t) = Cx(t) + Du(t)$.
`g=tf(b,a,T);`	Discrete time transfer function T $G(z)$ with $Y(z) = \frac{B(z)}{A(z)}U(z)$.
`g=tf(b,a);`	Continuous time transfer function $G(s)$ with $Y(s) = \frac{B(s)}{A(s)}U(s)$.
CONVERSIONS	
`gd=c2d(gc,Ts)`	Converts a continuous time system **gc** to a discrete time one **gd**.
`gc=c2d(gd)`	Converts a discrete time system **gd** to a continuous time one **gc**.
`gdnew=d2d(gd,Ts)`	Converts a discrete time system **gd** to another discrete time system **gdnew** with new sample interval **Ts** (the old one is implicit in **gd**.
SIMULATION	
`y=step(g);`	Step response
`y=impulse(g);`	Impulse response
`y=lsim(g,u,t);`	Simulation with general input **u**

Table 27.2 Control system toolbox (`help control`). Model definition and simulation.

CONTROL SYSTEM TOOLBOX AND THE LTI OBJECT, PART II	
GRAPHICAL SYSTEM ANALYSIS	
`nyquist(g)`	Plot of Nyquist curve
`bode(g)`	Plot of Bode diagram
`margin(g)`	Plot of Bode diagram with stability margins printed
`sigma(g)`	Plot of the amplitude curve of the Bode diagram for scalar systems, the singular values for multi-variable systems.
`rlocus(g)`	Plot of root locus
`rlocfind(g)`	Plot of root locus with user interaction
NUMERICAL SYSTEM ANALYSIS	
`pole(g)`	Poles of the system
`tzero(g)`	Zeros of the system
`ctrb(g)`	Controllability matrix
`obsv(g)`	Observability matrix
CONTROL DESIGN TOOLS	
`g=ss(g);`	The functions below apply to state space models
`k=acker(g.a,g.b,p);`	Uses Ackerman's formula to compute the state feedback $u = -Kx$ such that the poles of the closed loop system are the ones specified in vector p
`k=place(g.a,g.b,p);`	Uses pole placement to compute the state feedback $u = -Kx$ such that the poles of the closed loop system are the ones specified in vector p
`k=lqr(g.a,g.b,Q,R);`	Uses linear quadratic design for *continuous* time systems to compute the state feedback $u = -Kx$ such that $\int (x^T Q x + u^T R u)dt$ is minimized
`k=dlqr(g.a,g.b,Q,R);`	Uses linear quadratic design for *discrete* time systems to compute the state feedback $u = -Kx$ such that $\sum (x^T Q x + u^T R u)$ is minimized
`gc=ss(g.a-g.b*k,g.b,g.c,g,d);`	Computes the closed loop feedback system $\overset{+}{x}= Ax + B(r - Ku), \quad y = Cx + Du$. Here r is the reference signal, which is the new LTI input. The same call works for both continuous and discrete time ($\overset{+}{x}$ stands for either $\dot{x}(t)$ or $x(t+1)$.

Table 27.3 Control system toolbox (`help control`). Analysis and control design.

SYSTEM IDENTIFICATION TOOLBOX	
GUI	
`ident`	Interactive data pre-processing, system identification and validation (GUI and directory names are identical)
COVARIANCE FUNCTION	
`Rx=covf(x,kmax);`	Estimates the covariance function $R_x[k] = \mathrm{E}[x[n]x[n-k]]$, using $\hat{R}_x[k] = \frac{1}{N}\sum_n x[n]x[n-k]$, $k = 0,1,2,\ldots,k_{max}-1$. See also `xcov` in `signal`
`plot(0:kmax-1,Rx);`	Plots the estimate
SPECTRUM ESTIMATION	
`P=etfe(y);`	Periodogram, equal to `abs(fft(y)).^2`.
`P=etfe(y,gamma);`	Blackman-Tukey's estimate of the spectrum, defined as $\Phi(\omega) = DFT(R_x[k])$, with sampling frequency `fs` and time window width `gamma`. The estimate is basically a smoothed version of `abs(fft(x)).^2`. using time window width `gamma`. See also `psd` in `signal`
`P=spa(y,gamma);`	Blackman-Tukey's spectrum estimate using time window width `gamma`.
`bode(P);`	Plots the estimate
`th=ar(y,na);`	Estimate of AR model
`bode(th);`	Plots model-based spectrum estimate
MODEL ESTIMATION	
`yd=detrend(y);`	Remove mean and trend in the signal first.
`th=arx(y,na);`	Estimate AR model $y[k] = \frac{1}{A(q)}e[k]$
`th=arx([y u],[na nb nk]);`	Estimate ARX model $y[k] = \frac{B(q)}{A(q)}u[k] + \frac{1}{A(q)}e[k]$. Alternative model structures with corresponding estimators are implemented in `oe`, `armax`, `bj`, `pem`
`present(th)`	Presents the model object
`yhat=predict([y u],th,k);`	Predict the model k step ahead
`compare([y u],th,k)`	As above, but presents the result as a plot.
`resid([y u],th)`	Correlation analysis for model validation
`idsim([u e],th)`	Simulation of a model
ADAPTIVE FILTERING	
`theta=rarx([y u],...` `[na nb nk],adm,adg);`	Recursive estimation of ARX (or AR) model. The adaptive algorithm is either RLS `adm='ff';`, LMS `adm='ug';`, NLMS `adm='ng';` or the Kalman filter `adm='kf';`

Table 27.4 System identification toolbox (`help control`)

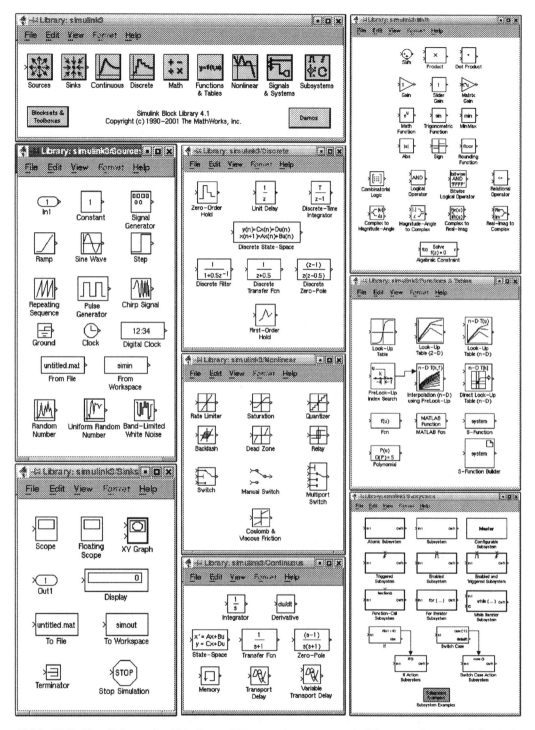

Table 27.5 Simulink standard blocksets. When typing `simulink` in Matlab, the upper left window appears on the screen. By double clicking on a specific blocks, the other windows are opened. Choose C-N or File-New-Model to open a blank window, and use the drag and drop principle. See Section 23 for an example. In Simulink version 4, there is a convenient tree-like structure for finding the blocksets.

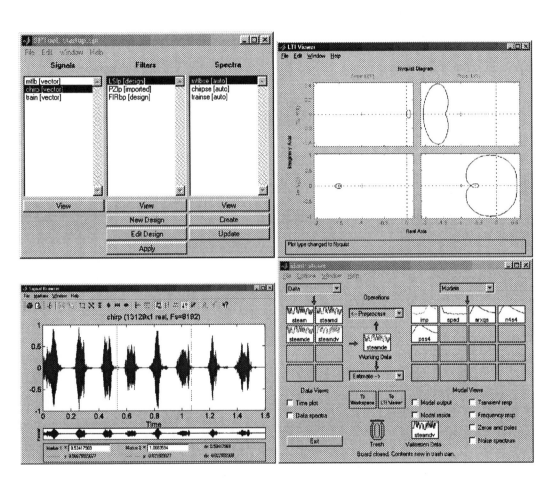

Table 27.6 Main GUI's of the toolboxes ident, control, signal.

Appendix E

GRAPHICS SUMMARY

Here we summarize the most important plot facilities. The plots are organized as follows:

- 2D graphics (**help graph2d**) illustrate vectors in general, with typical uses:
  - Short vectors for data visualization.
  - Long vectors for functions $y = f(x)$ and parametric two-dimensional functions $y = f_y(z), x = f_x(z)$ on the plane.
- 3D graphics (**help graph3d**) illustrate matrices in general, with typical uses:
  - Small matrices for data visualization.
  - Large matrices for functions $z = f(x, y)$.
  - Parametric two-dimensional functions $y = f_y(z), x = f_x(z)$.
- Images and graphical objects.

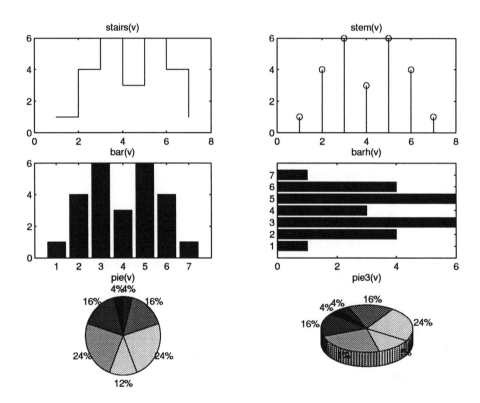

Figure 27.1 Different ways to visualize a short vector, here v=[1 4 6 3 6 4 1];

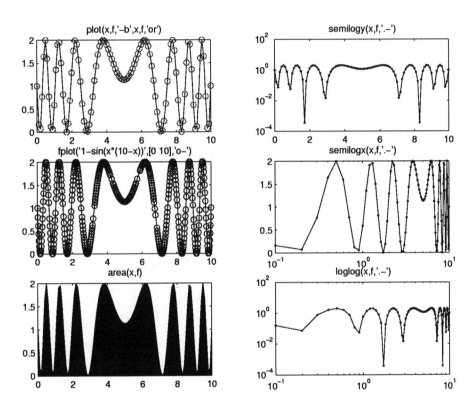

Figure 27.2 Different ways to visualize a long vector, here created as the function
x=0:0.1:10;
f=1-sin(x.*(10-x));

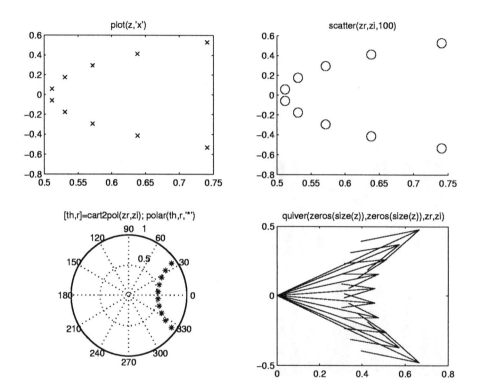

Figure 27.3 Different ways to visualize a vector of complex numbers, here created as
[b,a]=butter(10,0.2);
z=roots(a);
zr=real(z); zi=imag(z);

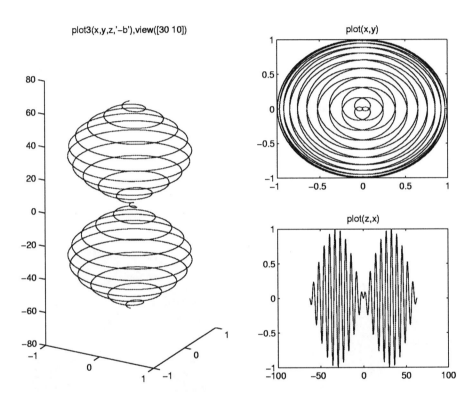

Figure 27.4 Different ways to visualize a two-dimensional parametric function, here created as
z=-20*pi:0.1:20*pi;
x=sin(0.05*z).*sin(z);
y=sin(0.05*z).*cos(z);

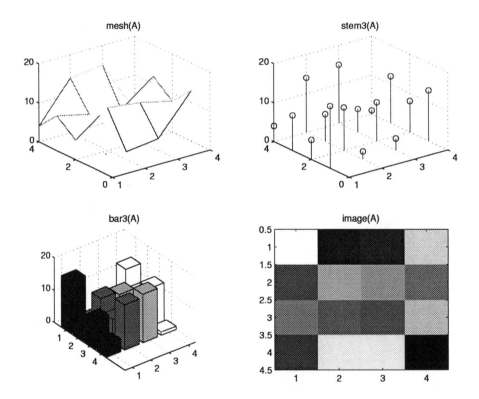

Figure 27.5 Different ways to visualize a small matrix, here created as A=magic(4);

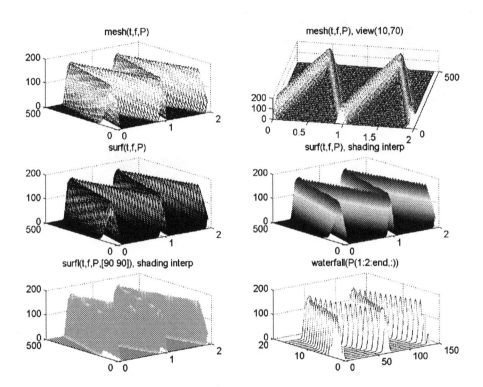

Figure 27.6 Different ways to visualize a large matrix in 3D, here created as
```
x=0:2000;
y=sin(2*pi*x.^2/2000);
[B,f,t]=specgram(y,64,1000,64,50);
P=filter2(ones(5),abs(B));
```

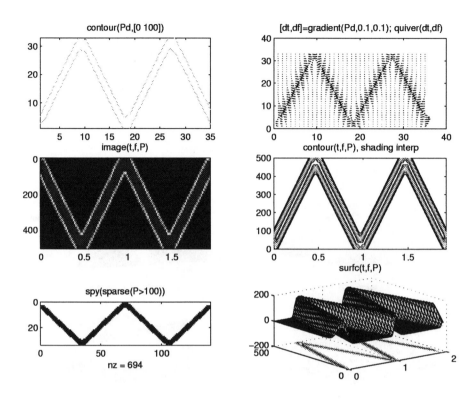

Figure 27.7 Different ways to visualize a large matrix in 2D/3D, here created as in Figure 27.6.

Figure 27.8 The patch is used to create a road sign.
```
patch([0.7 1 0.85],0.1+[0 0 0.3*sqrt(3)/2],'r');
patch([0.74 0.96 0.85],0.1+[0.02 0.02 0.3*sqrt(3)/2-0.04],'y');
t=0:30;y=sin(2*pi*t/20);
h=plot(0.8+t/300,0.03*y+0.17,'-k');
set(h,'LineWidth',12)
```

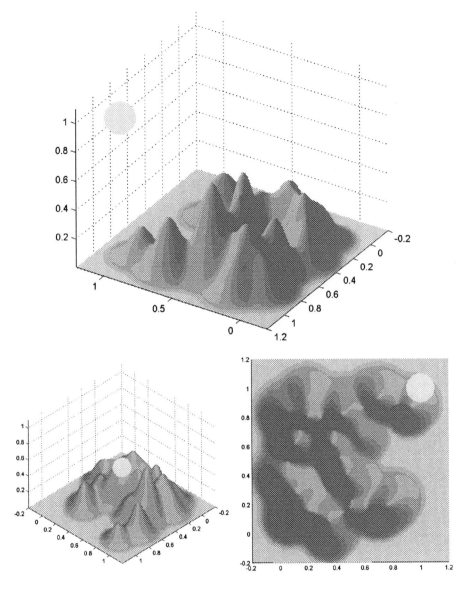

Figure 27.9 The landscape model is computed by
```
n=30; x=rand(n,1); y=rand(n,1);
[X,Y]=meshgrid(-0.2:0.01:1.2,-0.2:0.01:1.2);
Z=zeros(size(X));
for i=1:n;
Z=Z+0.2*exp(-100*(X-x(i)).^2-100*(Y-y(i)).^2);
end
surfl(X,Y,Z,[1 1 1]).
```
Here X,Y is the mesh grid, x,y are randmomized positions, and [1 1 1] is the light position used in surfl. The light is further illustrated by 'sun' generated by a surf plot of a sphere. The view angle is changed in the lower two plots to view([1 1 1]) (the light position) and view([0 90]) (from above).

BIBLIOGRAPHY

[1] The Mathworks. Learning MATLAB 6 – Student version. ISBN 0-9672195-3-1.

[2] Brian R. Hunt, Ronald, L. Lipsman and Jonathan M. Rosenberg. A Guide to MATLAB: for Beginners and Experienced Users. Cambridge University Press, 2001. ISBN 0521-00859-X.

[3] Edward B. Magrab, Shapour Azarm, Balakumar Balachandran, James Duncan, Keith Herold and Gregory Walsh. An Engineer's Guide to MATLAB. Prentice Hall, 2000. ISBN 0-13-011335-2.

[4] Andrew Knight. Basics of MATLAB and Beyond. CRC Press, Inc., 2000. ISBN 0-8493-2039-9.

[5] Jamal T. Manassah. Elementary Mathematical and Computational Tools for Electrical and Computer Engineers Using MATLAB. CRC Press, Inc., 2001. ISBN 0-8493-1080-6.

[6] Delores M. Etter. Engineering Problem Solving with MATLAB, 2e. Prentice Hall, 1997. ISBN 0-13-397688-2.

[7] Brian D. Hahn. Essential MATLAB for Scientists and Engineers, 3e. Pearson Education South Africa, 2002. ISBN 1-868-91143-8.

[8] Rudra Pratap. Getting Started with MATLAB 5: A Quick Introduction for Scientists and Engineers. Oxford University Press, 1999. ISBN 0-19-512947-4.

[9] Delores M. Etter, David C. Kuncicky and Douglas W. Hull. Introduction to MATLAB 6. Prentice Hall, 2002. ISBN 0-13-032845-6.

[10] William J. Palm III. Introduction to MATLAB 6 for Engineers: with 6.5 Update. McGraw-Hill, 2001. ISBN 0-07-283300-9.

[11] Delores M. Etter and David C. Kuncicky. Introduction to MATLAB for Engineers and Scientists. Prentice Hall, 1996. ISBN 0-13-519703-1.

[12] Duane C. Hanselman and Bruce Littlefield. Mastering MATLAB 6: A Comprehensive Tutorial and Reference. Prentice Hall, 2001. ISBN 0-13-019468-9.

[13] Adrian Biran and Moshe M.G. Breiner. MATLAB 6 for Engineers. Prentice Hall, 2002. ISBN 0-13-033631-9.

[14] Joe King. MATLAB 6 for Engineers: Hands-on Tutorial. R.T. Edwards, Inc., 2001. ISBN 1-93021-706-4.

[15] Stephen J. Chapman. MATLAB Programming for Engineers, 2e. Brooks/Cole Publishing. ISBN 0-534-39056-0.

[16] Brian Daku. M-Tutor: An Introduction to MATLAB. Prentice Hall Canada, 1999. ISBN 0-13-083396-7.

[17] Marc E. Herniter. Programming in MATLAB. Brooks/Cole Publishing. ISBN 0-534-36880-8.

[18] Darren Redfern and Colin Campbell. The MATLAB 5 Handbook. Springer-Verlag, 1998. ISBN 0-387-94200-9.

[19] Eva Pärt-Enander and Anders Sjöberg. The MATLAB 5 Handbook. Addison-Wesley, 1999. ISBN 0-201-39845-1.

INDEX

abs, 8
addpath, 39
Adobe Illustrator, 26
all, 12
alphabetical order, 70
angle, 8
animation, 167
ans, 6
any, 12
arrow keys, 4
at, 58
AVI, 169
AVI, 167
axis, 25

backslash, 119
\ backslash operator, 30
band-pass filter, 194
band-stop filter, 194
bar, 37, 128
bin2dec, 163
bit error rate, 162
bit rate, 164
block matrix inversion formula, 32
break, 52
butter, 149, 194

c2d, 134
call by reference, 85

callback, 80
case, 47
cat, 65
Cayley-Hamilton theorem, 31
cd, 38
ceil, 6
cell array, 17
channel coding, 163
char, 16, 93, 163
cheby1, 194
cheby2, 194
chirp, 151
chirp, 78
chirp signal, 78
clabel, 34
clear, 18
clf, 23
clock, 50
coin tossing, 130
collect, 191
: indexing operator, 9
: indexing operator, 12
color-map, 83
colorbar, 33, 150
colormap, 83
combine, 191
comet, 167, 168
comet3, 167, 168
comma separated list, 17

213